Like this bo
"Like" this book.

Help get the word out about *Word Up!*

Tweet it. Post it. Share it every which way. Tell your colleagues, your friends, your family. Your mom. Especially your mom.

All the *Word Up!* links—Twitter, Google+, Facebook, YouTube— await you in one convenient place:

http://howtowriteeverything.com/pass-the-word

Type that URL, or jump to it with one scan:

Everybody needs a good word.

Thank you.

Teaching?
Studying?

http://howtowriteeverything.com/exercises

Acclaim for Word Up!

"Informative and funny."

—PENNY J. BEEBE, RETIRED SENIOR LECTURER IN WRITING AT CORNELL UNIVERSITY

"I simply love the book. It gives me ideas about my own direction. Marcia was smart to write it as a series of essays rather than attempting to create a formal style guide. She has a unique voice. I'm glad she let us hear it."

—EDDIE VANARSDALL, AUTHOR OF *CONTENT INSOMNIA* BLOG

"I read...with interest and pleasure. Nice examples, and I especially liked the discussion of stylistic reasons why one would or would not choose to use verb particles...Marcia has a wonderfully engaging writing style that is quite refreshing."

—MURIEL R. SCHULZ, COAUTHOR OF *ANALYZING ENGLISH GRAMMAR*, ON EXCERPTS

"Inspirational."

—MARTHA BROCKENBROUGH, AUTHOR OF *THINGS THAT MAKE US [SIC]*, ON "WHAT BRAND R U?"

"Marcia Riefer Johnston has packed this volume with nuggets of wise advice. Any aspiring writer would be wise to mine them."

—JACK HART, AUTHOR OF *A WRITER'S COACH*

"Light, fun, and enjoyable to read. Marcia approaches grammar and style with a refreshing perspective. I wanted to make a checklist of principles to remember as I edit my own writing."

—TOM JOHNSON, AUTHOR OF *I'D RATHER BE WRITING* BLOG

"Why can't style guides explain writing like Marcia Riefer Johnston?"

—KEITH KMETT, CERTIFIED USABILITY ANALYST

"I'm only thirty-five pages in, but I started at midnight last night and do you know what I did the moment I woke up? Turned on my light and resumed reading. Marcia Riefer Johnston has an engaging, clear, and humorous style. I am enjoying every word—and rolling on the floor laughing. I feel myself becoming more powerful already!"
—CAROLYN KELLEY KLINGER, PRESIDENT OF THE SOCIETY FOR TECHNICAL COMMUNICATION, WASHINGTON DC CHAPTER

"I admire Marcia's clarity and thoroughness. Her examples are excellent in that they illustrate distinctions admirably. She is fun to read."
—THOMAS P. KLAMMER, COAUTHOR OF *ANALYZING ENGLISH GRAMMAR*, ON "YOU DON'T KNOW FROM PREPOSITIONS"

"Marcia expertly weaves together writing strategy and tactics. Whether you're new to writing or a pro seeking handy tips, this book has you covered."
—COLLEEN JONES, AUTHOR OF *CLOUT*

"You rarely get this kind of knowledge in such an engaging way. Read this book like a collection of short stories."
—RAHEL ANNE BAILIE, COAUTHOR OF *CONTENT STRATEGY*

"This book embodies the adage *practice what you preach*. Marcia's writing style is powerful in its own way: engaging, funny, instructive, and supportive. I've longed for a book on writing to recommend to clients and colleagues ... and this is it!"
—KRISTINA HALVORSON, AUTHOR OF *CONTENT STRATEGY FOR THE WEB*

"Where has this book been all my life? *Word Up!* is a must-have for anyone who writes anything, anywhere, anytime. I wish I had found it twenty-five years ago."
—MAXWELL HOFFMANN, PRODUCT EVANGELIST FOR TECH COMM SUITE AT ADOBE SYSTEMS

"Witty."

"Clever and clear, funny and wise. I'm a tech editor who wants to transition to fiction editing. Quotes from *Word Up!* will be more effective in my comments than those from some stuffy, grammar-geekish style guide (sorry, 'Chicago')."

"I recommend *Word Up!* to my students. I enjoy the casual approach and sense of humor infused in each chapter."

"I find myself forcing others to listen while I read 'this great part' out loud every few minutes. My dogs will soon be English experts! I plan to give this book to all my English teachers."

"Since reading Marcia's book, my sentences are 42 percent more powerful. Plus they have jazz hands."

"Like vitamin-enriched chocolate. This book brought me a smile as well as instruction."

"Buy this book."

"Seriously. Buy this book."

Word Up!

Word Up!

How to Write Powerful Sentences and Paragraphs

(And Everything You Build from Them)

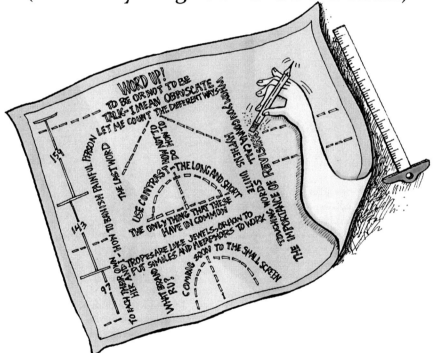

MARCIA RIEFER JOHNSTON

Northwest Brainstorms Publishing
Portland, Oregon

© 2012 by Marcia Riefer Johnston

Editing and proofreading: Ali McCart
Developmental editing and interior design: Vinnie Kinsella
Illustrations: Brian Hull
Logo, sentence diagram, and *seeing as/seeing that* drawing: Brian Poulsen

Northwest Brainstorms Publishing
Portland, Oregon

The chapters entitled "*To Be* or Not *To Be*" and "Talk—I Mean Obfuscate—Your Way
to the Top" appear, in different form, in the spring 1991 issue of *IABC Communicator*,
the newsletter of the Central New York chapter of the International Association of
Business Communicators.

21 20 19 18 17 16 15 14 13   1 2 3 4 5

ISBN: 978-0-9858203-0-5

Library of Congress Control Number: 2012943763

www.howtowriteeverything.com

For Brian and Elizabeth

A writer is somebody for whom writing is more difficult than it is for other people.
—THOMAS MANN, *ESSAYS OF THREE DECADES*

Contents

Foreword

Words are, of course, the most powerful drug used by mankind.
—Rudyard Kipling in a speech quoted in the *London Times*

If you're like me, you learned the basics of the English language from a well-intentioned adult. Someone like Mrs. White, my fifth-grade Language Arts teacher. Mrs. White knew her stuff. And she made sure her students did too. She used various memorization tools to ingrain spelling, grammar, and linguistics rules in our minds: *I before e except after c* and *Never end a sentence with a preposition.* Every day, as soon as the morning bell rang, she ran her students through a cavalcade of memorization exercises. Flash cards, chanting, rhyming, singing—nothing was off limits. Each student had to be able to recall, on demand, the rules of our mother tongue. No exceptions. Rules were rules.

Mrs. White's approach to teaching Language Arts worked (as far as she and other teachers like her were concerned). Her students mastered the rules. Our test scores proved it: her approach had succeeded.

Or had it? I can't speak for my classmates, but I had a difficult time turning this rote learning into effective writing.

By the time I reached university, the value placed on diagramming sentences and mastering semicolon usage had diminished. College professors emphasized critical thinking, problem solving, and communication. All that grammar, spelling, and linguistics stuff was pushed aside. Gone were the days when Mrs. White would award a gold star for stellar punctuation performances. The new rules we had to memorize in post-secondary education involved little of what Mrs. White and her ilk had drummed into us.

After college I took a job as a technical writer at an information-technology firm. My job involved writing, lots of it. I created e-mail newsletters, reports, proposals, presentations, manuals, configuration guides, online help, and training materials. Before they were published and delivered to customers, they had to pass muster with the editors. It was a strange world in which Mrs. White's rules resurfaced, colliding with the rules I had learned in college. Certain editors—well-intentioned but inconsistent rule enforcers—used their knowledge, our company style manual, and grammar rules from Mrs. White's era to impose their will. That approach to editing provided little value to those of us who wanted to grow as writers. We continued to make the same mistakes. Editors continued to correct them.

That was nearly twenty years ago. The teaching of writing has changed. Today, most schools in the United States are no longer required to teach cursive writing.[1] Education authorities in New Zealand have considered allowing students to use texting abbreviations in examinations.[2] *U mean i can rite liKe this IN a x-zam? KeWL. Way 2 go dude.*

As shocking as those developments may seem to those of us raised a generation or more ago, we can take heart in some of the changes in the way language skills are taught. Some teachers are de-emphasizing the memorization techniques that Mrs. White and others have used to present rules out of context—the drill-and-kill method—in favor of integrating grammar lessons into a broader study of reading and writing.[3] These teachers realize that their students, like all of us, have a better chance of becoming strong writers when rules are presented in a context of strong writing.

---

1. Scott Elliott, "New Standards Don't Require Students to Learn Cursive Writing," *Indianapolis Star*, July 9, 2011, http://www.indystar.com/article/20110710/NEWS04/107100364/New-standards-don-t-require-students-learn-cursive-writing.

2. NZPA and Mandy Smith, "Principals Oppose Text Language in Exams," *NZ Herald*, November 9, 2006, http://www.nzherald.co.nz/nz/news/article.cfm?c_id=1&objectid=10409902.

3. Matthew Tungate, "Grammar Is a Subject from Which Reading and Writing Should Not Be Removed," *Kentucky Teacher*, July 10, 2012, http://www.kentuckyteacher.org/features/2012/07/grammar-is-a-subject-from-which-reading-and-writing-should-not-be-removed.

That's why the book you're holding is so important. If you're like me (and I'm wagering you are), you're a good writer who wants to follow the rules, but every now and again you run into language situations that make you question whether your recollection of the rules is serving you well.

You're not alone.

Marcia Riefer Johnston's collection, *Word Up! How to Write Powerful Sentences and Paragraphs (And Everything You Build from Them)*, is loaded with practical advice for improving your writing by making good use of rules that matter. It does more than preach grammar. It helps you take command of words.

Each well-written lesson provides you with easy-to-remember tips for improving your prose. Johnston reveals interesting and peculiar facts about our language, including some that will delight you. She uses well-placed humor to demystify some often confused rules. She helps you decide when to abide by rules and when to break them. She looks at both sides of certain rules that even experts disagree on. You may be surprised to find that some rules aren't rules at all; they're guidelines that were intended to steer us in the right direction but may have done the opposite.

Word Up! is packed with assumption-obliterating advice that would likely earn praise from Shakespeare himself. It's a buffet of grammar and style snacks. Something for everyone. Take what you want. Leave the rest for the next reader.

If Mrs. White were still teaching today, I'm certain she'd use this book. She knew the power of words. Despite my inability to master all the rules, she would be glad to know that she instilled in me a desire to better understand our language and to wield it with authority and confidence. She'd see this book, whether used in the classroom or beyond, as a means of achieving those goals.

Now, what are you waiting for? Turn the page. Let's get started.

—Scott Abel, The Content Wrangler

Preface

Most of the chapters in this book grew out of entries posted on my blog, *Word Power*,[4] named after a high-school class that I had expected to dislike. The teacher, Larry Wray, introduced himself as a lover of words. What a strange idea! He then handed out yellow workbooks entitled *Word Power: A Short Guide to Vocabulary and Spelling* by Dr. Byron H. Gibson, who made some outrageous claims of his own:

> "Words are power!"

> "Teacher, your students will come back through the years to thank you for giving them this help in their single most important objective, learning words and learning them accurately, on which all other life objectives depend."

> "This guide has been prepared to be the single most helpful book you have ever studied."

Hyperbole! I might have thought, had I thought in quadrisyllabic words. I liked writing well enough. I kept a journal. I valued self-expression. (My dad was a psychiatric social worker.) But I used whatever words came to me, undiscerningly, the way a hitchhiker hops into the first car that stops to offer a ride. Along comes Dr. Gibson, telling me that all life objectives depend on the accurate learning of words. Right.

---

4. Marcia Riefer Johnston, *Word Power* blog, http://marciarieferjohnston .wordpress.com. I have migrated that blog's contents to a new website, http://how towriteeverything.com, which includes a blog along with information about this book.

I suspended my disbelief. I did the work. This class was supposed to help us prepare for the SATs, after all. My classmates and I, following what the author called the Gibson-Gordis method, learned Latin and Greek prefixes, roots, and suffixes. For example, we memorized prefixes (*a-, amphi-, ana-, anti-, apo-, cata-*) and associated them with words (*a*moral, *an*emia, *amphi*bious, *amphi*theater). We filled out worksheet after worksheet.

The SATs came and went. I flew to Europe to live for a year with an Austrian family. I forgot about Mr. Wray and Drs. Gibson and Gordis. But my separation from the only language that came naturally to me sharpened my awareness of the importance of words. For months, I struggled constantly to communicate in German, coming up against my linguistic limitations in every interaction. Deprived of familiar words, I realized how much I had taken them for granted. With the awe of a child realizing that her parents had once had childhoods, I came to see that words, in any language, had lives of their own—long histories, complex genealogies—that I could only guess at.

I discovered, for example, that *shoe* had taken centuries to become *shoe*. Through those same hundreds of years, the nearly identical but more resonant *Schuh* had evolved to require more space between the tongue and teeth. (Both words ostensibly descend from the Proto-Germanic *skōhaz*, which in the Iron Age meant "covering."[5]) Similarly, I connected *father* and *Vater*, which must have derived from the same original—or should I say *ur*?—mouth movements.

During that year in Austria, the phrase *Es fällt mir ein*—"It falls into me," literally, or "It occurs to me"—became one of my favorites. Every day, all sorts of new understandings fell into me. Language fell into me.

When I came home, I moved on to college. I read Homer and Hemingway. I read like I had never read before. And I wrote. I grabbed every writing opportunity that presented itself, on campus and off.

I became a lover of words.

Since then, I've done more kinds of writing than you want to hear about. Each has taught me something about words and their ability to

---

5. "Shoe," *Wiktionary*, last modified June 21, 2012, http://en.wiktionary.org/wiki/shoe.

instruct, console, uplift, devastate, tickle, bore, confuse, and persuade. Writing leads me to insight and satisfaction. It deepens my relationships. It brings me pleasure. It earns me a decent wage.

When I set up my blog, the need to name it brought Mr. Wray to mind for the first time in decades. It fell into me that no name would do but *Word Power*.

That yellow workbook? Not the single most helpful book I ever studied. But Dr. Gibson predicted correctly when he said, "Teacher, your students will come back through the years to thank you." Several years before I thought to thank him, Larry Wray died. Thanked or not, he must have known the value of what he taught us.

> During that year in Austria, the phrase *Es fällt mir ein* — "It falls into me," literally, or "It occurs to me" — became one of my favorites. Every day, all sorts of new understandings fell into me.

Word power. I aim to pass it on.

Acknowledgments

I wish I could name everyone who has nurtured or inspired my writing. I have learned from more teachers, readers, writers, and word lovers than I can count.

Thanks to Joe Wycoff and the other team-teachers at Chesterton High School who seared the keyhole essay structure into your students' minds. That odd keyhole shape, which you drew over and over on the chalkboard, saw me—and eventually my own students—through many a term paper. You taught me the importance of structure.

Thanks to whoever donated that Hemingway book to the library of my Austrian high school. Helga Gruber, it had to be you. When, for the first time in my life, I hungered for English, Hemingway fed me.

Thanks to Russ Tutterow at Lake Forest College for opening the stage to me when I took a notion to write plays and for patiently coaching me on all things drama.

Thanks to Rosemary Cowler at Lake Forest College for bringing even the dustiest literature to pulsing life for all who had the good fortune to sit in your classroom. I'll never forget your delivery of this *Beowulf* line in Old English: "WAY-och under WOCH-num, WAY-oh THIN-dum THA!"

Thanks to the mentors, whose names I wish I could remember, who said yes when I was looking for summer jobs that would pay me to write, including one job at a small-town newspaper and another at a big-city public-relations firm.

Thanks to Ray Carver and Toby Wolff, who saw enough promise in my short stories to accept me into the Syracuse University graduate writing program. You made us, your lucky students, feel that our words mattered.

Thanks to Jean Howard at Syracuse University for guiding me through an exhilarating summer of intensive work on that *Shakespeare Quarterly* article. You have set my standard for collaborative revising (re-vising!).

Thanks to Karen Szymanski, who had the vision to start a technical-writing internship all those years ago at Magnavox CATV and who had the heart to let both applicants share in it. That internship kicked off a long career for me. Until you came along, I had never heard of tech writing, and I had never dreamed that I could make a living as a writer.

Thanks to Don Flynn and Bob Odell, who hired me at Genigraphics for my first tech-writing job, taught me about coneheads and pinheads, and showed me what a blast people can have at work.

Thanks to Anne Coffey and Sally Cutler, who hired me at Word-Wrights (despite my omitting some hyphens in that proofreading test) and who taught me how to run a writing consultancy.

Thanks to Lori Lathrop, whose workshop introduced me to the exquisite complexities of book indexing.

Thanks to all who volunteered feedback on some part of this book's content or on the process of publishing. Your encouragement kept me going. The book has more to offer because of you: Melissa Amos, Rahel Bailie, Mark Baker, Lydia Beck, Penny Beebe, Aleks Bennett, Dave Bennett, Martha Brockenbrough, Andrea Carlisle, Bill Chance, Howard Collins, Sandee Craig, Kim Delaney, Angela Della Volpe, Karen Dunn, Doug Eaton, Bryan Garner, Elisabeth Grabner, Joe Gollner, Aaron Gray, Kristina Halvorson, Jack Hart, Adrienne Hartz, Mark Hartz, Catherine Hibbard, Jen Jobart, Diana Johnson, Tom Johnson, Colleen Jones, Tom Klammer, Carolyn Kelley Klinger, Keith Kmett, Jane Laysen, Deb Lockwood, Mary Lou Mansfield, Gwyn Mauritz, Pam Minster, Michelle Morie-Bebel, John Morrison, Melinda Musser, Art Plotnik, Becca Pollard, Ginny Redish, Amy Reyes, Lisa Riefer, Kathy Sage, Muriel Schulz, Connie Shank, K. Vee. Shanker, Laurie Sherer Simon, Keith Spillett, Amy Spring, Francis Storr, Eddie VanArsdall, Gee Gee Walker, Cheryl White.

Thanks to Erica Caridio for reviewing this book's indexes and sharing your gentle wisdom.

Thanks to Cheryl Landes for helping me understand the limitations of today's e-book indexing options and for working toward better options for tomorrow's writers. Your generous, insightful review of this book's indexes made a difference. You called attention to some blind spots. What a great feeling to know that I have none left.

Thanks to Jan Wright for the additional insights that helped make the indexes work better both for technology and for readers. I appreciate all that you, too, are doing to improve tomorrow's e-book indexes.

Thanks to Scott Abel, the—*The*—Content Wrangler, for writing my foreword and for sharing your enthusiasm and sense of humor with the wide world of fellow wranglers.

Thanks to the pros whose expertise and cheer helped me navigate the many stages of developing this book: Jessica Glenn (of Jessica Glenn Book Publicity) for your boundless book-shepherding abilities and your unparalleled connections; Tina Granzo (of City Beautiful Design) for your impressive skill and unfailing responsiveness in creating the book's website; Brian Hull (of BriAnimations Living Entertainment) for illustrating the book with such flair; Vinnie Kinsella (of Indigo Editing & Publications) for your insights on the text and your careful design work; Ali McCart (of Indigo Editing & Publications) for your editing acumen and your scrupulous attention to detail.

Thanks to the "lackadaisicals" for the eight-year conversation about books and life. Every author dreams of readers (and friends) as astute and passionate as you: Karen Baum, Teresa Craighead, Carrie Koplinka-Loehr, Tracy Mitrano, Rebecca Nelson, Sue Rakow, Christina Stark.

Thanks to Curt and Martha Johnston for being who you are and for raising a son who loves words. Without him (which is to say without you), I might have written a book, but I could not have written this book.

Thanks to Shannon Wood for your faith in this book through its two-year odyssey and for your lifelong faith in me.

Thanks to my sister, Wendy Hood, for commenting on my every blog post. You enrich my writing almost as much as you enrich my life.

Thanks to my dad, Dennis Riefer, who always kept books around and who believed that I could do anything.

Thanks to my mom, Stella Robertson, the clearest explainer I know. No wonder I gravitate toward how-to writing. You taught me how to knit and how to pack a suitcase. You taught me, and are still teaching me, how to live.

Thanks to my daughter, Elizabeth Poulsen, for your careful reading of early versions of these chapters. Your insights opened new possibilities for the book. You've been opening new worlds to me since the day you were born.

Thanks to my son, Brian Poulsen, for the skill and professionalism you brought to the two line drawings and the propositionally dense logo (and thanks for teaching me about propositional density). You've been inspiring me since the day you were born.

Thanks to my husband, Ray Johnston, for sustenance of every kind, for rescuing this book from my attempts at sports references, for laughing in all the right places, for knowing when to say, "It's not done," and for finally saying, "It's done."

Introduction

For months, I brainstormed titles for this book. Several times, I thought I had The One. Too long, people said. Too dull. Too this. Too that. My title would find *me*, my editor promised. Then it happened. I was reading a company newsletter, of all things, when I came across a bit of hip-hop slang: *Word*. I heard echoes of my son and his friends. *Word*. The ultimate in concise affirmation. The verbal equivalent of extending your fingertips—just the tips—to slip someone some skin.

A moment later, as predicted, the full title found me. I practically heard it, as if someone had said it in my ear. *Word Up*. It felt quirkily perfect. Snappy yet laid back. A name with cachet. A name with street cred. It seemed to say, "Hey. You. Interested in words? Got some writing to do? This book is for you."

You may write blog posts, e-books, e-mails, executive summaries, e-zine articles, hospital-hallway signs, presentations, proposals, lab reports, letters to the editor, love letters, lunch-bag notes, movie reviews, news stories, novels, online help, plays, poems, proposals, recipes, reference manuals, scholarly critiques, speeches, term papers, tweets, user-interface text, video scripts, web pages, or white papers. You may write for a million readers or for one. You may use a pen, a typewriter, a wiki, or an XML-authoring tool. You may be a grammar snob, or you may think that "grammar snobs are great big meanies."[6] You may write because something within you says you can't not write—or because your boss says you can't not write. No matter what you write, or how or why, you and I and every other writer have two things in common: we use words, and we want someone to want to read them.

---

6. Credit for this phrase goes to June Casagrande, author of *Grammar Snobs Are Great Big Meanies: A Guide to Language for Fun and Spite* (New York: Penguin, 2006).

How do you get people to want to read your words? Know your subject. Know your audience. And write powerfully.

This book can help you write powerfully.

Another book on writing! Doesn't the world already have too many writing books? Maybe. If you piled up all the books on writing, you'd have a precarious, weird-looking stack reaching…way up there. But the world can't have too many writing books of the kind I like to read, the kind I set out to write. This book doesn't say the same old things in the same old ways. This book follows its own advice. Practices what it preaches. Shows what it tells. This book uses powerful writing to talk about powerful writing.

Powerful writing is writing of any kind that accomplishes something, that gets through, that works. It's a handshake between friends, an instant of connection and understanding. You can find powerful writing in reports, résumés, refrigerator notes, and in any other artifact of communication in which the words of a thoughtful writer meet the interests of a receptive reader. Powerful writing entertains, heals, motivates, sells, enlightens. It marks the biggest and smallest occasions of human existence. Powerful writing changes things—for a person, a classroom, a country, a planet.

Powerful writing often accompanies other powerful elements of communication. This book doesn't talk about those elements. It doesn't talk about the power of images, video, audio, or any other type of content that writers may need to think about. Multimedia elements are sometimes worth thousands of words. But multimedia elements don't say it all. And they don't make writing work.

This book also says little about the power of technology. Servers and search engines push words out into the world in mighty ways. Computers enable writers to tag, tweet, and Google Translate. The Internet gives writing reach and reusability—*technopower*, as content strategist Rahel Anne Bailie calls it.[7] Technology amplifies writing. But technology doesn't make writing worth finding. It doesn't make writing work.

---

7. Rahel Anne Bailie, "Defining Content in the Age of Technology," *Intentional Design Inc.* blog, October 18, 2011, http://intentionaldesign.ca/2011/10/18/defining-content-in-the-age-of-technology.

Finally, this book doesn't talk much about the power wielded by professionals like content strategists, information architects, content-management-system experts, taxonomists, user-experience designers, informaticians, and other "info-slingers with a very wide range of titles."[8] People in these evolving fields help organizations manage (and help consumers navigate) massive websites, sprawling knowledge bases, and other immeasurable labyrinths of information—including, of course, zillions of words. (Kristina Halvorson, author of *Content Strategy for the Web*, puts words at the center of strategists' attention: "Most often I talk about content as text" because "text is everywhere...text is different...text is messy as hell."[9]) The pros who do this heroic, strategic work determine what gets written, why it's needed, what form it takes, who creates it, what metadata gets assigned to it, who governs it, who has access to it, where it's stored, which devices and applications it's compatible with, how and when and by whom it's organized and labeled and compiled and tested and delivered. These folks determine whether the content is technologically intelligent[10] and adaptive.[11] They decide which bits of information to update and which to send off to that great archive in the sky. People in these challenging roles dazzle with their ability to align writing goals with the goals of a company and its customers. They tie writing efforts to the

---

8. Erin Kissane, *The Elements of Content Strategy* (New York: A Book Apart, 2011), 34.

9. Kristina Halvorson, *Content Strategy for the Web* (Berkeley: New Riders, 2010), ix–x.

10. Visionaries Ann Rockley and Charles Cooper define *intelligent content* as "content that is structurally rich and semantically categorized, and is therefore automatically discoverable, reusable, reconfigurable, and adaptable." (*Managing Enterprise Content: A Unified Content Strategy*, 2nd ed. [Berkeley: New Riders, 2012], 16.) In that book's foreword, author Kristina Halvorson gives her own definition: "content that's free from the constraints of a document or page, and therefore free to adapt to any context or platform" so that it can be delivered to "the right people, in the right place, and at the right time." For my book review, see "'Managing Enterprise Content' 2nd Ed.: A Book Review," *How to Write Everything* blog, May 13, 2012, http://howtowriteeverything.com/managing-enterprise-content-2nd-ed-a-book-review.

11. Adaptive content is information that changes as needed—without human intervention. An adaptive instruction might show up as *click* on a laptop, *tap* on a tablet, and *say select* in a car's GPS. Rockley and Cooper define *adaptive content* as content that "automatically adjusts to different environments and device capabilities to deliver the best possible customer experience" (*Managing Enterprise Content*, 134).

bottom line. But they don't, for the most part, do the writing. They don't make writing work.

Skilled writers make writing work.

To make writing work, you need tools: universal, timeless, genre-transcending tools referred to, collectively, as command of the language. Command of the language! There's no app for that.

Command of the language requires, for starters, an understanding of grammar. This book is not about grammar, but it touches on grammar, sometimes briefly and sometimes in depth. Grammar, I've been surprised to discover, is not a fixed set of labels, definitions, and rules but an evolving body of knowledge full of subtleties and differences of opinion. The more you know about grammar in all its complexity—including the innumerable types of words that words can be and the dauntingly rich variety of ways that they can work in sentences—the more intentionally, confidently, accurately, and powerfully you can wield words, and the more easily your readers can make sense of them. (Grammar aficionados will especially appreciate the assumption-obliterating chapters "You Don't Know From Prepositions," page 49, and "A Modern Take (Is *Take* a Noun?) on Parts of Speech," page 61, as well as the revelatory Glossary, page 191. I make this claim because, while developing these sections, I came to appreciate the obliteration of some of my own assumptions and the revelations of some new terms and definitions.)

> **Command of the language! There's no app for that.**

To acquire command of the language, you need to learn not only about grammar but also about usage, style, linguistics, rhetoric, and more. How do you learn it all? For starters, read and read and read. Embrace your inner wordie. Devour a dictionary or two. Bone up on your bdelygmia (as in "You're a foul one, Mr. Grinch, you're a nasty wasty skunk").[12] Ruminate

---

12. Richard Nordquist, "Bdelygmia: The Perfect Rant," *About.com Grammar & Composition* blog, 2012, http://grammar.about.com/od/rhetoricstyle/a/Bdelygmia Rant07.htm. (Don't fall for my faux alliteration; the *b* in bdelygmia—"de-LIG-me-uh"—is silent.)

on rhetoric, "the art of influence."[13] Grab some advice from Grammar Girl.[14] Get your Garner on.[15] Take a writing workshop. Read up on language usage—and on everything else. Read menus. Read subway posters. Read graffiti. Read Mark Strand, Maeve Binchy, William Shakespeare.

Also listen … to lyrics, to TV shows, to movies, to podcasts, to your neighbors.

Most of all, write, and write, and write. Plop down words over and over, exchange them, move them around. Savor their sounds. Take them apart—roots, suffixes, prefixes—and put them back together. Build sentences. From those sentences, build paragraphs. Arrange those paragraphs into all kinds of sturdy, useful, beautiful, glorious—and not-so-glorious—constructions.

Whatever you read, hear, or write, notice what works and what doesn't—for you and for others. Everything you read, everything you hear, everything you write has something to teach you. Gradually, you take command. Command of your language.

If this book helps you along the way, please tell others about it. Some of the people you talk with every day would love to write more powerfully. Why not say to them, "Hey. You're interested in words. You've got writing to do. I have just the book for you."

Word Up!

---

13. Colleen Jones, *Clout: The Art and Science of Influential Web Content* (Berkeley: New Riders, 2011), 43–79.

14. Mignon Fogarty, *Grammar Girl: Quick and Dirty Tips for Better Writing* blog, http://grammar.quickanddirtytips.com.

15. If you love language and sometimes run into seemingly unanswerable questions about its usage—that is to say, if you love language—you want this book: Bryan A. Garner, *Garner's Modern American Usage*, 3rd ed. (Oxford: Oxford University Press, 2009). For a daily taste, subscribe to Garner's Usage Tip of the Day via e-mail: http://www.oup.com/us/subscriptions/subscribe/subscriptions.jspx?view=usa.

Up with (Thoughtful) Prescriptivism

When the original version of my now-quite-altered essay "To Each Their Own" (page 27) appeared as a guest post on Tom Johnson's *I'd Rather Be Writing* blog, it stirred up some controversy.[16] Readers' comments prompted me to address some questions: Who am I—who is anyone—to give advice on writing or speaking? Why bother talking about how language should be used, given that language keeps changing no matter what?

Language, the argument goes, simply is.

Academics call this point of view descriptivism. Descriptivists claim to make no judgments: English speakers say what they say, no right or wrong about it. As one commenter wrote in response to my post, "Grammar is value neutral ... People have many ideas about how language *should* work, but that's roughly equivalent to having ideas about how the weather should work." (This analogy—*just as efforts to influence weather are futile, so too, by extension, are all efforts to influence language*—illustrates the flaw in analogous reasoning. Weather and language may both be huge and impossible to control on the grand scale, but the similarity ends there. Unlike weather, communication doesn't just happen. And unlike observers of weather—who never need to make clouds—observers of language, like their fellow language users, need to communicate.)

On the other side of the debate, we have prescriptivists, those who prescribe. They recommend using words this way vs. that way. Their

---

16. To review the discussion, see my guest post on Tom Johnson's *I'd Rather Be Writing* blog, April 16, 2011, http://idratherbewriting.com/2011/04/16/guest-post-to-each-their-own.

opinions conflict, of course. Some opinions are more nuanced or better researched than others. Because we can turn to no absolute authority for arbitration, disagreements get heated.

Here's one way to sum up the difference. Prescriptivists compare *should*s and pick one. Descriptivists tell people they should stop saying *should*.

The descriptivist-prescriptivist debate has raged (not too strong a word) for generations. Bryan Garner—a man who calls himself a "descriptive prescriber,"[17] a man whose Twitter bio (as of this writing) says, "Fall in love with language & it will love you back"—describes the debate in two thorough and thoroughly engaging essays: "Making Peace in the Language Wars" and "The Ongoing Struggles of Garlic-Hangers."[18] For my purposes, the salient point, as Garner puts it, is this: "Literate people continue to yearn for guidance on linguistic questions."[19] Writers and editors need help "solving editorial predicaments."[20]

> Unlike observers of weather—who never need to make clouds—observers of language, like their fellow language users, need to communicate.

Even after a quarter century of professional writing, I still yearn for linguistic guidance, and I still struggle with editorial predicaments. I'm grateful to my fellow prescriptivists—all those creators of style guides, writers of grammar books, newspaper columnists, essayists, English teachers, editors, and other opinion wielders who've had the audacity to share their insights on language. The rightness or wrongness of their pronouncements is beside the point. These folks have helped me strengthen my writing, understand how language works, make decisions. They have helped me convey meaning. I want to do the same for others.

---

17. Garner, *Garner's Modern American Usage*, xl, xliv.
18. Ibid., xxxvii–lx.
19. Ibid., xlvii.
20. Ibid., xviii.

So—despite the judgments of those nonjudgmental descriptivists who say things like "Prescriptivism must die!"[21]—I prescribe.

I have no illusions about my ability to stop change. As one blog commenter noted, "You can't fight the language. *They* and *their* are changing, and no essay is going to stop that." True. In case you haven't figured out that language changes—or in case you'd simply enjoy an entertaining snapshot of the ever-evolving English language—pick up (you'll need both hands) the latest edition of *Garner's Modern American Usage*, and flip it open to any page. Throughout this tome, Garner applies what he calls his Language-Change Index to indicate "how widely accepted various linguistic innovations have become."[22] This index describes five stages of acceptance, from Stage 1, "Rejected," to Stage 5, "Fully accepted." For example, *enormity*, which today usually means "enormousness," once widely meant "hideousness."[23] Garner ranks today's usage—"*enormity* misused for *immensity*"—at Stage 4: "Ubiquitous but...opposed on cogent grounds by a few linguistic stalwarts (die-hard snoots)."[24]

Clearly, language changes. Still, I see value in writing about conventions and practices and opinions that may help people—those looking for help—to succeed in this difficult business of communicating.

How do we prescriptivists decide which conventions and practices and opinions to prescribe? We study rules and guidelines. We consider context. We discern, research, compare, weigh, wonder. We love the language.

Descriptivists love language too. Mark Twain, a descriptivist in practice if not in title, captured colloquialisms as accurately and passionately

---

21. Google this phrase at your own risk.

22. Garner, *Garner's Modern American Usage*, xi.

23. Ibid., 306–307.

24. Ibid., xxxv. If the word *snoot* all by itself makes you smile, by all means treat yourself to Garner's delicious discussion of the term (756). If you already know what Garner means by *snoot*—David Foster Wallace's family's acronym for *syntax nudnik of our time* or *Sprachgefühl necessitates our ongoing tendance*—you probably also know yourself, happily, to be one. And you probably forgive me for loving footnotes almost as much as Wallace did.

as any scholar. Only a descriptivist could have penned lines like these, spoken by Tom Sawyer about Aunt Polly: "She never licks anybody—whacks 'em over the head with her thimble—and who cares for that, I'd like to know. She talks awful, but talk don't hurt—anyways it don't if she don't cry."[25] Twain's nonjudgmental rendering of everyday speech, his unique brand of descriptivism, left a mark on world literature.

Descriptivism and prescriptivism don't rule each other out. *Up with prescriptivism* doesn't mean down with anything. You down with that?

---

25. Mark Twain, *The Adventures of Tom Sawyer*, illustr. ed., (Hartford: American Publishing Company, 1881), 27.

PART I

Up with Words

The bad or indifferent English to be met with in private and business correspondence and in a good deal of printed matter is often due not so much to gross mistakes in grammar or the use of words as to poor craftsmanship, that is, to sheer clumsiness in using the tool of language.
—M. ALDERTON PINK, *CRAFTSMANSHIP IN WRITING*

To Be or Not To Be

A verb, Senator, we need a verb!
—*Doonesbury* comic strip

Want one tip, a single bloat-busting strategy guaranteed to energize your sentences? Dump *to be*. Wherever you spy a weak, static, insubstantial *be*-verb—*be, being, been, am, are, is, was, were, have been, could be, will be, won't be*—think, *Opportunity*.

We can't call every *be*-verb weak. A *be*-verb works plenty hard when it acts as an auxiliary—especially when it works with the main verb to pack a wallop (*We are busting the habit of using weak* be-*verbs*) or to convey a colloquialism (*We be stylin'* or *I'll tell you what I'm up to if you tell me what you're on about, and then I must be off* [26]). A *be*-verb also pulls its weight when it points to existence itself (*We write, therefore we are*). Pow!

I'm not talking about strong *be*-verbs like those.

I'm talking about "flabby *be*-verbs,"[27] *be*-verbs lacking in muscle, *be*-verbs that powerful writers hunt down and expunge. Specifically, I'm talking about *be*-verbs that act as linking verbs (*The house is beige*), as expletive-supporting verbs (*It's a beige house*), or as passive-voice auxiliaries (*The house was painted beige*). Ho-hum. We'll get into fuller descriptions of these three types in a minute. First, so that you can see what I mean, let's look at some examples of sentences transformed by tossing the flaccid *be*-verbs.

---

26. This example contains three phrasal *be*-verbs: *be up to* ("be doing"), *be on about* (in the United Kingdom, "be discussing"), and *be off* ("depart").

27. Garner, *Garner's Modern American Usage*, 612.

Before: "Education's purpose is to replace an empty mind with an open one." (Malcolm S. Forbes)

After: Education replaces an empty mind with an open one.

Before: "Nothing is more revealing than movement." (Martha Graham)

After: Nothing reveals like movement.

Before: "A scheme of which every part promises delight can never be successful." (Jane Austen)

After: A scheme of which every part promises delight can never succeed.

Before: "In all pointed sentences, some degree of accuracy must be sacrificed to conciseness." (Samuel Johnson)

After: All pointed sentences must sacrifice some degree of accuracy to conciseness.

Before: Our product is better than your product. (any company)

After: Our product eats your product's lunch.

Before: "There are known knowns. These are things we know that we know. There are known unknowns. That is to say, there are things that we know we don't know. But there are also unknown unknowns. There are things we don't know we don't know." (attributed to a one-time US secretary of defense)

After: We don't know diddly—except to avoid *there are.*

The limitations of *be*-verbs have intrigued people at least as far back as the 1930s, when Alfred Korzybski developed the discipline of "general semantics." Among other things, Korzybski explored what he called the "structural limitations" of these verbs.[28] His teachings inspired a

---

28. "Alfred Korzybski," *Wikipedia*, last modified May 28, 2012, http://en.wikipedia .org/wiki/Alfred_Korzybski.

student, D. David Bourland Jr., to develop E-Prime (English-Prime, also denoted E ′), a form of English that excludes *be*-verbs. E-Prime rejects statements like *This painting is beautiful*, which presents judgment as fact, in favor of statements that "communicate the speaker's experience," such as *I like this painting*.[29]

Okay, you get it. Weak *be*-verbs: who needs them!

How do you spot a weak *be*-verb? Let's get back to the three types: the linking verb, the expletive-supporting verb, and the passive-voice auxiliary.

A *be*-verb that acts as a linking verb usually robs your sentence of power. A linking verb (usually but not always a *be*-verb[30]) creates "an equivalency"[31] between a subject and its complement.[32] It acts as a simple pass-through, an equal sign. Take the statement *Their faces are pale from all this grammar talk*. The *are* acts as an equal sign: *faces = pale*. This verb connects (links) the two words: end of story. With few exceptions, a linking-verb sentence benefits when you punch it up by replacing the weak verb with a speeding bullet of a verb: *Their faces blanch with all this grammar talk*. Your readers win.

Similarly, you can ditch *there is, there are, it is, it was*, and other phrases formed by a *be*-verb plus an expletive. Here, *expletive* means not an obscenity but a "dummy word," like *there* or *it*, that has no grammatical function. When you diagram an expletive sentence, the expletive floats above the other words like a let-go balloon—a disconnected, puffed-up nothing. Of sentences that start with *it is important*

---

29. "E-Prime," *Wikipedia*, last modified June 13, 2012, http://en.wikipedia.org/wiki/E-Prime.

30. Linking verbs include not only *be*-verbs but any other verb—*seem, appear, become, remain, grow, get*—that acts as an equal sign in a given sentence. Does your face grow pale from all this grammar talk?

31. Jack Hart, *Storycraft: The Complete Guide to Writing Narrative Nonfiction* (Chicago: University of Chicago Press, 2011), 113.

32. Defining *complement* would take a footnote longer than even I can see burdening you with. I'd have to start by saying that a complement is a word or phrase that completes the sense of a subject, an object, or a verb, and then I'd have to define the definition, and then we'd have to get into the types of complements (like adverbial complements, adjectival subject complements, nominal subject complements...and those are just the types of linking-verb complements), so let's not even start, okay?

to note that or *it is interesting to note that*, Bryan Garner says, "These sentence nonstarters merely gather lint."[33] (If you're afflicted, as I am, with the need to mark up your books, you probably just highlighted *nonstarters* and *lint*.) Instead of saying *It is important to tighten your sentences*, say *Tighten your sentences*. Your readers win.

> When you diagram an expletive sentence, the expletive floats above the other words like a let-go balloon—a disconnected, puffed-up nothing.

Finally, a passive-voice auxiliary sucks power from a sentence almost every time. Passive voice is a verb form that shows the subject receiving the action instead of performing it. Tighten and strengthen such sentences by converting passive voice (*The blood was drained from my face by all that grammar talk*) to active voice (*All that grammar talk drained the blood from my face*). Your readers win.

You won't have an easy time of it, eradicating all these types of weak *be*-verbs from your writing. They pop constantly to mind as you form thoughts. You can't suppress them. So don't. Let them flow. As you generate ideas—as you create your drafts, as you brainstorm, as you think inventively—let the weak *be*'s be. Later, when you hone, zero in on these verbs and on the revising opportunities they represent.

In some cases, weak *be*'s merit keeping. They can enable you to do the following:

Linking-verb *be*'s:

Play with a common expression	Boring is in the eye of the beholder.
Create a cadence	"The play's the thing." (Shakespeare)
Position a key word at the end	Rules are for breaking.
Position a key word at the beginning	Upside-down was, in fact, how Samuel felt.

---

33. Garner, *Garner's Modern American Usage*, 486.

Define a term	An essay is a short piece of writing on one topic.
Emphasize a classification	Yes, this chapter is an essay.
Emphasize an equation	"The medium is the message." (Marshall McLuhan.)
Emphasize a metaphor	"Love is a rose." (Neil Young)
Heighten the diction level	"May the Force be with you." (*Star Wars*)
Add umph	Now *that's* what I'm talking about.

Expletive-supporting *be*'s:

Fill the meter in a line of poetry	"Something there is that doesn't love a wall." (Robert Frost)
Evoke melodrama	"It is I, Snidely Whiplash!" (*Dudley Do-Right*)

Passive-voice *be*'s:

State an action of an unknown doer	The wheel was invented around 8000 BC.
Avoid naming a known doer	Shakespeare was born on April 23, 1564.

(For the sake of these examples, let's assume that Shakespeare was Shakespeare. You might count yourself among those who believe that the works of Shakespeare were penned by someone else, possibly a group of playwrights who were almost certainly not all born on April 23, 1564. I offer that statement about Shakespeare's birth, and all these follow-on statements in the parentheses, to reinforce the point that sometimes the reader is better served by what I call weak *be*'s. For example, although converting passive voice to active voice usually improves a sentence, nothing would be gained by saying, "Mary Arden gave birth to Shakespeare on April 23." If you want to talk about Shakespeare, don't make his mother the subject of your sentence—even when she is, for once, at least grammatically, the actor.)

Clearly, *be*'s—even weak *be*'s—need to be. But you'll use fewer and fewer of them as you fortify your writing. Make the break! The difficulty may surprise you. Stick with it; persistence will reward you. You'll discover the satisfactions of writing more intentionally. You'll use fewer clichés, fewer adjectives, fewer adverbs, fewer nouns, fewer...words. As you wean yourself off weak *be*'s, you'll use more—and more forceful—verbs, the strongest part of speech there is. The strongest part of speech. Period.

> **If you want to talk about Shakespeare, don't make his mother the subject of your sentence—even when she is, for once, at least grammatically, the actor.**

Talk—I Mean Obfuscate—
Your Way to the Top

> *"Synergy" is one of the key words used by business professionals to indicate that they have no clue as to what business they are actually in.*
>
> —DAVE BARRY, "IDIOT'S GUIDE TO ENGLESH," MR. LANGUAGE PERSON

You've tried dressing like a boss and acting like a boss. Still waiting for that promotion? Try talking like a boss. Why let all that bureaucratese go to waste?

For starters, don't raise problems. Raise concerns. "I have a problem" makes you sound like a whiner. "I have a concern," on the other hand, sets you up as a responsible corporate citizen. Open this way, and then state your problem. People take you more seriously when you use fuzzier language. (Notice how seriously you're taking me right now.)

Is your department overworked? You don't need people. You need resources. Better yet, you have a resource concern. As often as possible, use two words instead of one, and you'll impress your audience twice as much.

Need an extra chair? You could get one. Then again you could procure one. Extra syllables get you extra stuff. I mean, extra *matériel*.

In the middle of a meeting, don't blurt out, "Have you lost your mind?" Say, smoothly, "Let's take that offline."

Don't tell people what to do; give them action items. Don't make plans; negotiate logistics. Don't prepare; do legwork. Don't get people to agree with you; get them to buy in. First, though, triangulate (don't bounce) your ideas off them.

Don't buy cheap goods. Buy cost-effective goods. And don't spend money. Spend monies. Even better: appropriated monies (monies from someone else's budget).

Above all, don't ask for a raise. Discuss a salary action.

When managers talk, listen. Gerard Braud, author of *Don't Talk to the Media Until ...*, did just this, gathering favorite business buzzwords from three hundred people. His list, which he strung together into a speech, landed him on stage at the International Association of Business Communicators 2012 World Conference. Borrow from Braud, whose speech you'll find on YouTube.[34] Borrow from everyone. Practice your new words on the way to and from work. Say them out loud. Roll them around in your mouth. Ease them into your conversations. At first they'll feel artificial, euphemistic, inflated, and repetitive. And redundant. Don't worry. You will get comfortable with your new language.

> "I have a problem" makes you sound like a whiner. "I have a concern," on the other hand, sets you up as a responsible corporate citizen. Open this way, and then state your problem.

And your new peers.

Be prepared for new challenges in your managerial role. Imagine catching the CEO in the elevator. You tell her that you have a critical-path concern about resources. You're not asking for sign-off yet. You have a few ideas to triangulate first to ensure a cost-effective resolution. You offer to take an action item to do the legwork and negotiate the logistics if she'll appropriate the monies. You'd even postpone your salary action if doing so would procure her approval.

You've got this game down.

She says, "Let's take that offline."

---

34. Gerard Braud, "The Worst Speech in the World," YouTube video, 6:41, from a presentation at the International Association of Business Communicators 2012 World Conference in Chicago, posted by IABCtv, June 27, 2012, http://www.youtube.com/watch?v=9LIAI2tEApc&feature=player_embedded.

The Only Thing That These Signs Have in Common

The only poor decisions are the ones you don't follow through on.
—YOGI BERRA, COMMENCEMENT ADDRESS AT SAINT LOUIS UNIVERSITY

I'm up to something here, only I'm not saying what. I'll give you only one hint: you don't want to get caught putting a certain word in the wrong place.

Okay, I'll give you more than one hint, only you have to find the others on your own.

To confirm the answer, skip ahead—but only after you've studied these examples, all taken from signs I've seen.

ALCOHOL!

I ONLY DRINK TO MAKE **YOU** MORE INTERESTING!

THINK QUALITY!

QUALITY ONLY HAPPENS WHEN **YOU** CARE TO DO YOUR BEST!

These signs have only one thing in common: the *onlies* come too early. *Only*, which has been called "perhaps the most frequently misplaced of all English words,"[35] belongs next to the word or phrase it limits. Compare the originals with these rewrites:

- I drink only to make you more interesting. (The original — "I only drink..." — implies, "Drinking is the only thing I do to make you more interesting.")

35. Garner, *Garner's Modern American Usage*, 592.

- Quality happens only when you care enough to do your best. (The original— "Quality only happens ..."— implies, "The only thing that quality does when you do your best is *happen*; it doesn't do a dang thing more.")

- I have a kitchen only because it came with the house. (The original—"I only have a kitchen..."—implies, "I don't do anything else with that kitchen but *have* it. *Have have have*, all day long.")

- I can please only one person a day. (The original—"I can only please one person a day..."—implies, "All I can do is please that person every day. It must be dull, getting nothing but pleased. 'Pleased today, pleased tomorrow...can't a person get anything but pleased around here?'")

- Change available only with postal transaction. (The original—"Change only available..."—implies, "Sorry, folks, all of our postal-transaction change is the available kind. If you want unavailable change, go find another post office.")

- Casket for sale. Used only once. (The original—"Only used..."— implies, "Nothing else was done to this casket. It was used, okay? Just used. Only used. That's all you need to know.")

It surprised me to learn that English speakers have been slopping their *onlies* around for hundreds of years. Scottish rhetorician Hugh Blair noted in the eighteenth century that "with respect to such adverbs as *only*, *wholly*, *at least*, and the rest of that tribe...we acquire a habit of throwing them in loosely." Blair made allowances for this looseness in conversation, but he held writers to a higher standard: "In writing, where a man speaks to the eye...he ought to be more accurate."[36]

Speak to the eye. Useful advice. And not for *only* only.

---

36. Hugh Blair, *Lectures on Rhetoric and Belles Lettres*, vol. 1 (1783; 11th ed. 1809), 245, quoted in Garner, *Garner's Modern American Usage*, 593.

Her and I:
How to Banish Painful Personal-Pronoun Pairings

I never made a mistake in grammar but one in my life and as soon as I done it I seen it.
—CARL SANDBURG IN *A DICTIONARY OF LITERARY QUOTATIONS*

My father is living with my wife and I.

A businessman sent this statement out to thousands of readers. Does the *I* hurt your ears? If it doesn't—if the *I* sounds right to you, or if it sounds funny but you aren't sure why, or if you never know whether to say *I* or *me* but you favor *I* because you've heard lots of otherwise well-informed people talk that way—you're not alone. Pronoun misuse saturates American parlance.

The trouble arises in sentences that involve two parties. No one would say, "My father is living with I." What trips people up is the *and*.

So get rid of it, if only for a moment. Cover the *and* with your mind's hand before you speak or write.

Example:

Him/He and *me/I* went fishing this morning.

Cover up the *and*. Look at each pronoun by itself:

Him/He went fishing this morning.
Me/I went fishing this morning.

No problem. No one would say *Him went fishing* or *Me went fishing*. Don't let that little troublemaker, *and*, change a thing. If it's *He went fishing* and *I went fishing*, then it's *He and I went fishing*.

Every time.

If your ear needs recalibrating, try these sentences. Say the correct versions out loud. Repeat until what is right sounds right. (Note that the *I/me* and *we/us* choices come last. As Bonnie Trenga, author of *The Curious Case of the Misplaced Modifier*, says, putting first-person pronouns last is "the polite thing to do. 'Me first' is a bad attitude in life, and so it is in grammar, too."[37])

He/Him and *I/me* went to the store to get ice cream.	He and I went to the store to get ice cream.
The armchair was big enough for *her/she* and *I/me*.	The armchair was big enough for her and me.
Are you coming to the game with *she/her* and *me/I*?	Are you coming to the game with her and me?
Her/she and *I/me* will drive you home.	She and I will drive you home.
That hybrid truck is perfect for *she/her* and *me/I*.	That hybrid truck is perfect for her and me.
Throw the football to *her/she* and *I/me*.	Throw the football to her and me.
Build *him/he* and *I/me* a house.	Build him and me a house.
Want to hear about *he/his* and *my/I's* plan?	Want to hear about his and my plan?
The bleachers have plenty of empty spots for *they/them* and *us/we* to sit comfortably.	The bleachers have plenty of empty spots for them and us to sit comfortably.
With practice, *him/he* and *I/me* learned new grammar habits.	With practice, he and I learned new grammar habits.

---

37. Bonnie Trenga, "Between Me and You?" *Grammar Girl* blog, March 27, 2009, http://grammar.quickanddirtytips.com/between-me-and-you.aspx.

To Each Their Own

"Look, your worship," said Sancho; "what we see there are not giants but windmills, and what seem to be their arms are the sails that turned by the wind make the millstone go."

"It is easy to see," replied Don Quixote, "that thou art not used to this business of adventures."

—MIGUEL DE CERVANTES, THE INGENIOUS GENTLEMAN, DON QUIXOTE OF LA MANCHA

They has finally gone too far.

You may think I'm denouncing the singular *they*, as in sentences like these:

Open the profile of a *friend*, and add *their* phone number so it's easy to call *them*.

Health management allows *one* to take care of *themselves*.

As the *lover* seeks *their* beloved, so must you focus on what you want.

If you think I'm talking about this coupling of plural pronouns (*their, them, themselves*) with singular nouns (*friend, one, lover*), you're partly right. I do avoid the singular *they*—even though people have used it for centuries and even though many style guides condone it. But when I say, "*They* has finally gone too far," I'm talking about a recent trend. I'm talking about computer-generated sentences like these:

Jane wants to add you to *their* network.

Jim has updated *their* profile.

Oh, Jane. Oh, Jim. You have been neutered!

Even usage authority Bryan Garner, who allows that the singular—indeterminate—*they* "promises to be the ultimate solution" to the pronoun problem,[38] says, "*John got their coat* is ghastly."[39]

What are writers to do? Shall we train ourselves to shrug instead of flinching? Shall we adopt this usage ourselves as modern and inevitable?

I understand this unfortunate unpluraling of pronouns. English fails us here. It offers no word for "his-or-her." We have no *lui*, which those lucky French can say when they mean "to him or her." In Grammar Girl Mignon Fogarty's words, "English has a big, gaping pronoun hole."[40] None of our singular third-person pronouns—*he, his, him, himself, she, hers, her, herself*—stands in adequately for *person* or *anyone* or *each*. We have only a handful of singulars, each in some way lacking.

> *He:* "To each *his* own" conveys an old-fashioned gender bias.
>
> *She:* "To each *her* own" reverses the bias.
>
> *S/he:* "To each *his/her* own," when spoken, requires a hand motion.
>
> *He or she:* "To each *his or her* own" works, but few choose this option, perhaps because it strikes many as "awkward."[41]

People who reject these imperfect choices fill the need by co-opting the conveniently gender-neutral, if inconveniently plural, *they*, as in "To each *their* own."

(When I published the original version of this essay as a guest post on Tom Johnson's *I'd Rather Be Writing* blog on April 16, 2011, I assumed that my title, "To Each Their Own," would give people a jolt. The jolt was on me when I picked up a magazine several weeks later

---

38. Garner, *Garner's Modern American Usage*, 739–740.

39. Bryan A. Garner, e-mail message to the author, April 26, 2011.

40. Mignon Fogarty, "Generic Singular Pronouns," *Grammar Girl* blog, October 20, 2011, http://grammar.quickanddirtytips.com/he-they-generic-personal-pronoun.aspx.

41. A listener named Betty, quoted in Fogarty, "Generic Singular Pronouns."

and discovered a full-page Honda ad with the headline "To Each Their Own" in large, 3-D letters.[42])

I'm not caving. I continue to finesse "the pronoun problem" by writing around it. For example, the phrase *As the lover seeks their beloved* lends itself to any of the following alternatives:

Turn singulars into plurals	As *lovers* seek *their* beloveds ...
Go ahead, use *his or her*	As the *lover* seeks *his or her* beloved ...
Switch occasionally between feminine and masculine	*lover ... his; lover ... her*
Switch to a direct address: *you*	*Lover*, as you seek *your* beloved ...
Switch to the more inclusive *we*	As we *lovers* seek *our* beloveds ...
Remove the pronoun altogether	As the *lover* seeks *the* beloved ...

Why not apply the same techniques to automated phrases? Why script "[Name] has updated their profile" when an alternative like "[Name] has an updated profile" lies so close at hand?

Alas, the day approacheth fast wherein the singular *they* shall pain the ear of humankind no more. The battle—sayest thou else?—is all but lost. Still, I make this final plea.[43] ~~A person must stand their ground.~~ Let us stand this ground together. Fight with me!

---

42. For a critique of this ad's headline, see Richard Read, "2012 Honda Civic Comes with Ninjas, Zombies, Grammar Problems," *MotorAuthority* blog, April 26, 2011, http://www.motorauthority.com/news/1058971_2012-honda-civic-comes-with -ninjas-zombies-grammar-problems.

43. Diction level (the degree of formality indicated by word choice and phrasing) communicates in its own way beyond the definitions of the words. Typically, you don't want diction level to call attention to itself, but, like any other tool, sometimes it enables you to accomplish something—emphasize a point, let your voice come through, have a little fun—that you can't do as well any other way.

Whom Ya Gonna Call?

If there's somethin' strange
in your neighborhood
Who ya gonna call?
　　—The Rasmus, *Ghostbusters*

Whom. You can't say the word without sounding snooty. As soon as your lips close on the uncool *m*, your nose tilts up.

Imagine a group of rockers walking out on stage, announcing themselves as (watch their noses) The Whom. Visualize Dr. Seuss sitting at his typewriter, writing about (again the nose) all the Whoms in Whomville. Picture Abbott and Costello standing at their microphones doing Whom's on First.

Silly, I know. The point is that *whom*, the word itself—right or wrong—offends some people's sensibilities.

"Who's she calling *offended*?" I can practically hear people whispering.[44] Even talking about the word *whom* feels somehow impolite. Presumptuous. Un-American. Dropping the *m* has become a form of cultural sensitivity, an expression of democratic values, a way of saying, "We're in this together." If you and I were created equal, common usage seems to say, why shouldn't *who* and *whom* be equal too?

But *who* and *whom* are no more interchangeable than you and I. Ignoring this truth (which is apparently not held to be self-evident) doesn't make it less true.

How do you know which term is correct? More to the point for the *whom*-averse, when is it safe to use *who*?

---

44. If you wonder about the *who* in "Who's she calling offended?" come back to this question when you finish reading this chapter.

Try this. In the split second before you say *who*, think *he*. If *he* works, *who* works. But if your *he* needs the *m* in *him*, then your *who* needs an *m* too.

Think of it this way:

who = he

(Both pronouns are in the nominative case.)

whom = him

(Both pronouns are in the objective case.)

Example:

You want to ask this: *Who* did you walk with?

You do the *he* test: *He* did you walk with?

You flip the words around for more natural phrasing: Did you walk with *he*? (Ugh.)

You swap in *him*: Did you walk with *him*? (Yes.)

You realize you could ask this: *Whom* did you walk with?

Or this: With *whom* did you walk?

With practice, your brain flies through these steps. You simply know.

Who cares? Often no one. Take Twitter. How many tweeple do you suppose complain about the phrase *Who to Follow* in their menu bar? This gaffe probably bothers only a fraction of the millions of people who use this site every day.

Hold on. A fraction of millions. That could be a lot of bothered people.

No one says that you have to use the *m* word. If you don't want to, don't. George Thorogood would never have hit the charts with a song called "Whom Do You Love?" But think before you use *who* as a substitute. Many people know the difference. Who knows when one of them is listening?

Hyphens Unite!

It pains me to say this, but I may be getting too mature for details.
—JERRY SEINFELD, *SEINFELD*

My friend Mark notes that hyphens seem to be disappearing. "Not sure why," he says. "Hyphens make reading easier." He's talking about those times when two or more adjectives join forces, working together as a compound adjective (also known as a phrasal adjective or unit modifier) in front of a noun. One of my favorite examples comes from a sign not far from my neighborhood. In large letters, it gives this command: *Fear Free Dentistry.* Maybe these dentists intend to scare people away from free dentistry. Probably, though, they intend to advertise fear-free dentistry. (Their omission scares me away. I don't want anyone that sloppy coming at me with a drill.)

Does the lowly hyphen—that dinky half-dash, that barely-there conjoiner of words, that "pest of the punctuation family"[45]—deserve a whole essay in a book on writing powerfully? Is any punctuation mark less emblematic of power? If you were choosing teammates, you'd pick the hyphen last. A hyphen doesn't even merit sand in the face; bullies simply ignore it, inflicting the ultimate humiliation: leaving it out.

But when you see the hyphen for what it is, when you take the time to appreciate its unique qualities, you'll find it a powerful ally indeed.

Example:

true blue friend

---

45. Sophie C. Hadida, *Your Telltale English* (rev. ed. 1942), 133, quoted in Garner, *Garner's Modern American Usage*, 679.

Do you need a hyphen here? Try this test: say each adjective (*true* and *blue*) with the noun separately. *True friend.* That makes sense. *Blue friend.* That makes sense only if you're talking about a Smurf. So you don't have a true blue friend. *True* and *blue* work together. Call on the hyphen's unifying force, and you've got a true-blue friend.

Try the test on *clean energy consultant.* Pair each adjective with *consultant* separately. *Energy consultant.* That almost makes sense. *Clean consultant.* That makes sense only if the consultant just took a shower. *Clean* and *energy* work together: clean-energy consultant.

Other hyphenless headscratchers:

> sick ward nurse (a ward nurse with the flu)
>
> light green suitcase (a green suitcase that weighs little)
>
> ride on mower (a ride on a mower)
>
> little used cigar (eww!)

Usually, people can decipher a phrase like this from its context—after they stop, go back, and reread the words. But why make them reread? Why slow them down when a hyphen could speed them along? For example, let's say you run a prestigious hospital, and you're about to print a full-page ad on the back cover of the *New York Times Magazine* with this headline (complete with these unfortunate line breaks):

> A Father
> Son Bond So Close,
> They're Joined At
> The Liver.

You'd want to stop the press and unite *father* and *son* (father-son bond...) rather than force people—millions of people in this case—to stop and reread a headline that so inappropriately separates this dad from this boy.[46]

---

46. This hyphenless headline, which splashed across the back cover of the *New York Times Magazine* on November 27, 2011, also suffers from a noun-pronoun mismatch. Grammatically, the *they* refers to the would-be subject, *bond*, as if to say, "The bond

Should you always hyphenate a compound adjective (that is, two or more words working as one adjective) when those words come right before a noun? Some say yes. Commonly, though, when a whole phrase, noun and all, becomes widely recognized, the hyphen disappears. For example, even in the language-usage-curmudgeon-filled-technical-writing world, the hyphen has all but dropped out of certain common terms, like *content management system* or (more controversially in the curmudgeonliest circles) *quick reference card*. Those comfortable with these omissions argue that, in these cases, the hyphen no longer has a job to do.

In support of the dehyphenation of frequently used phrases, writer Edward Johnson notes that

> In large letters, the sign gives this command: *Fear Free Dentistry.*

"whereas *science-fiction writer* would normally be hyphenated, in a work that used the compound constantly it would not be: *Science fiction has changed since the days of early science fiction writers Jules Verne and H. G. Wells.*"[47]

I rarely omit the hyphen, even in frequently used phrases. Sometimes a style guide (or a boss) tells me to leave it out in certain contexts, and so I do. Otherwise, though, if the hyphen's knack for uniting could prevent even a few readers from stumbling, why not send the little guy in? As Johnson goes on to say, "Even the most familiar compounds can be ambiguous, and the writer, who knows the intended meaning, often will not notice the ambiguity; only the reader will."[48] *The Chicago Manual of Style* says, "With the exception of proper nouns (such as *United States*) and compounds formed by an adverb ending in *ly* plus an adjective … it is never incorrect to hyphenate adjectival compounds before a noun."[49] Usage authority Bryan Garner states the risk of going

are joined at the liver." Ouch again. Not the best way for a hospital to advertise its attention to detail.

47. Edward D. Johnson, *The Handbook of Good English: Revised and Updated* (New York: Facts on File, 1991), 205.

48. Ibid., 205.

49. *The Chicago Manual of Style*, 16th ed. (Chicago: University of Chicago Press, 2010), 373.

hyphenless this way: "almost all sentences with unhyphenated phrasal adjectives will be misread by *someone*."[50]

By all means, hyphenate when you find yourself in compounds-comprising-more-than-two-words situations—or better yet, rewrite!

Try this exercise in empathy. Read the following sentence in slow motion, as if you were sliding a strip of paper across the words to the right, revealing one word at a time.

> Marie's dad swears that next December he will avoid the last minute shopping frenzy.

You reach *avoid the last* with no trouble. Then *minute*. Split-second pause. Did Marie's dad want to avoid the last minute? You read on: *last minute shopping*. Aha! You back up and mentally insert a hyphen: *last-minute shopping*. There. Now, onward again: *frenzy*. Wait. Yet another hyphenless space—this time the one between *minute* and *shopping*—has stopped us. Did this dad want to avoid only the last-minute kind of shopping frenzy? Maybe next year he plans to hit the earlier shopping frenzies? How can we read on? How can we make sense of it all? We need another hyphen to come to the rescue: *last-minute-shopping frenzy*. Ahh. Reason is restored.

Save your readers from such distress with a little help from an under-celebrated hero. Hyphens, unite!

---

50. Garner, *Garner's Modern American Usage*, 627.

To Hyphenate or Not To Hyphenate After a Noun: That Is the Wrong Question

Asking questions is more important than finding answers—why?
—TOM JOHNSON, *I'D RATHER BE WRITING* BLOG

This job is long-term.

This job is long term.

Do you need the hyphen here? Most authorities say no. Don't hyphenate a compound modifier when it follows the modified noun. Before the noun, yes (*This is a long-term job*), but after, no (*This job is long term*).

Most authorities also point out exceptions. They say that some compounds need a hyphen even when they follow the noun. Which compounds, though ... *razor-sharp*? *risk-averse*? *time-sensitive*? *all-encompassing*? *cost-effective*? *blue-green*? Authorities disagree. Some defer to dictionaries, but you can't necessarily go by a dictionary. As *The Chicago Manual of Style* says, "When such compounds *follow* the noun they modify, hyphenation is usually unnecessary, even for adjectival compounds that are hyphenated in *Webster's* (such as *well-read* or *ill-humored*)."[51]

Good luck figuring out *blue-green* vs. *blue green*, for example. According to *Chicago*, compound adjectives formed with color words are "hyphenated before but not after a noun."[52] On the other hand, Edward D. Johnson, author of *The Handbook of Good English*, says,

---

51. *The Chicago Manual of Style*, 373–374.
52. Ibid., 375.

when it comes to "noun + noun color compounds such as *blue-green*" following the noun, "I advise retaining the hyphen."[53]

Let's look at one more can't-win example: *cost-effective* vs. *cost effective*. Do we hyphenate this compound after a noun? Jane Watson, who calls herself "North America's Grammar Guru," says no. She would have us write, *This program is cost effective.*[54] Just as definitively, the Case Western Reserve University Division of Student Affairs says yes. Their style guide would have us write, *This method is cost-effective.*[55] Crazy-making! As John Benbow, once editor of the Oxford University Press stylebook, is widely quoted as warning, "If you take hyphens seriously, you will surely go mad."[56]

So much for seeking the right answer.

Happily, I'm seeking not a right answer but a right question. Most authorities don't tell you that if you wonder, *Do I need a hyphen here?* after the modified noun, you ask the wrong question. They don't tell you what you most need to know: that a post-noun modifier almost always follows a *be*-verb (*is, are, was*) or some other linking verb (*seem, appear, become, remain, grow, get*). And they don't tell you that linking verbs almost always signal an opportunity to strengthen a sentence.

So what question should you ask yourself when faced, heaven forbid, with sentences like these?

This job is long term.

That child is razor-sharp.

The suit is blue-green (or blue green).

---

53. Johnson, *The Handbook of Good English*, 204.

54. Jane Watson, "Hyphens with Adjectives," *BizWritingTip* blog, March 13, 2012, http://bizwritingtip.com/?p=2952.

55. "Commonly Mis-hyphenated Words," Student Affairs IT Operations Group Website Management writing style guide, Case Western Reserve University, 2012, http://studentaffairs.case.edu/support/web/style/hyphenwords.html.

56. John Benbow, *Manuscript & Proof: The Preparation of Manuscript for the Printer and the Handling of the Proofs*, (New York: Oxford University Press, 1937), 92.

Ask yourself, *What do I have to say about that long-term job, that razor-sharp child, that blue-green suit?* Then, eliminate the linking verb (as described in "*To Be* or Not *To Be*" on page 13), and swap in some substance, some muscle:

> This long-term job pays more than anyone in Joan's family has ever made.
>
> Those razor-sharp kids speak twelve languages.
>
> Donovan thought that the blue-green suit made the professor look glamorous.

Then, maybe, you'll have yourself a sentence worth reading.

Let Me Count
the—Different?—Ways

The secret of good writing is to strip every sentence to its cleanest components.
—WILLIAM ZINSSER, *ON WRITING WELL*

I have nothing against the word *different*. Sometimes nothing else will do. For example, you offer me a bite of headcheese made from an old family recipe. "My," I might say. "That's...different."

But if you're stating or implying a number—twelve or a couple or a whole bunch—stop right there. Whatever countable things (plural) you're talking about—bridges or sheep or charities—don't call them different.

Consider these phrases:

> A dozen different bridges
>
> Two different sheep
>
> A lot of different charities

Those bridges had better be different, or something's awry with the laws of physics. Same goes for the sheep, even if one's a clone of the other. And if you're calling the charities different, don't ask me for a donation. If you so cavalierly waste words, why should I trust you with money?

The Pen Is Mightier Than the Shovel

Once the grammar has been learnt, writing is simply talking on paper and in time learning what not to say.
—BERYL BAINBRIDGE IN *CONTEMPORARY NOVELISTS*

Inuits can't have more words for snow than Upstate New Yorkers do. Having lived through twenty-some Upstate winters, I have my own words for snow. *Go away.*

I admit, though, that this white (or gray or black) stuff has its uses. For example, it inspires metaphorical thinking. One minute I'm chiseling frozen slush off the sidewalk, anticipating the moment when my son, Brian, and his high-school pals make their way up the soon-to-be-walkable concrete steps, and the next minute I'm thinking, *This is like editing.* To edit is to hack, hack, hack at the bits and heaps and chunks of junk obstructing the mind's way until either (a) we give up and leave our readers, like unfortunate pedestrians on a precarious trail, to fend for themselves or (b) we stand back in sweaty awe of the path we've created.

Is it coincidence that passages of text are called passages?

If you're hardy enough to apply a shovel to your own writing, give the heave-ho to the following words.

-ly words (and other vapid adverbs)	Many adverbs, especially the ones that end with *-ly*, ~~actually, truly, frankly, extremely, definitely, totally, really, simply, literally, basically~~ have less substance than the weightless, drifting snow that Inuits call weightless, drifting snow. ~~Relentlessly~~ chip away adverbs that ~~redundantly~~ repeat the meaning of the verb.

Don't jettison every *-ly* word, though. What's not to love about, say, a fabulously frumpy winter coat? As bestselling writer Arthur Plotnik points out, "joltingly fresh adverbs...are among the hottest locutions in contemporary prose."[a]

very, such, so
(and other empty intensifiers)

Despite their reputation as intensifiers, *very*, *such*, and *so* fail to intensify. When you come across one of these wimpy so-called intensifiers, hurl it into the bushes, and put a brawny verb to the heavy lifting. *I have a ~~very~~ strong desire to clear... I have ~~such~~ a strong desire to clear ... I ~~soooooo~~ want to clear... I long to clear this walkway.*

For those times when you need to make a point that's bigger than big—when even a hyperbolic Batman of a verb needs a Robin of an intensifier—sweep past all those nonboostingly stale booster words and pluck yourself a "wallopingly fresh superlative."[b]

Alternatively, let data do the intensifying: *For twelve sleepless hours, I've thought of nothing but clearing this walkway.*

not, no
(and other negative words)

Negative words often ~~do not have~~ **lack** muscle. They come in handy, though, as differentiators. The two *nots* add clarity here:

In the sentence Call me a shoveling fool, *the word* shoveling *is a verb only in form, not in function. It functions not as a verb but as an adjective.*[c]

---

a. Arthur Plotnik, *Spunk & Bite: A Writer's Guide to Bold, Contemporary Style* (New York: Random House, 2007), 37. I review this inspiring book here: http://howtowriteeverything.com/spunk-and-bite-arthur-plotnik-book-review.

b. This hyperbolically entertaining (I hope) Batman-and-Robin reference serves as an example of a superlative-freshening technique. For more, see Arthur Plotnik, *Better than Great: A Plenitudinous Compendium of Wallopingly Fresh Superlatives* (Berkeley: Cleis Press, 2011). Even if you have no immediate need for a wallopingly fresh superlative, you'll find this book a wallopingly good read.

c. For an introduction to a controversial (and valuable) modern way of looking at the forms and functions of words, see "A Modern Take (Is *Take* a Noun?) on Parts of Speech" on page 61.

the fact that	Does ~~the fact that you're sitting comfortably~~ **your comfort** have you feeling guilty about others doing all the shoveling?
just	~~Just~~ say no to *just*. *Just* ~~just~~ gets in the way. Sometimes it ~~just~~ slips out, especially right before a verb where it ~~just~~ adds no value. It ~~just~~ happens. ~~Just~~ forgive yourself, ~~just~~ delete it, and ~~just~~ move on.
begin to, start to	Don't ~~begin to~~ describe the beginning of an action. Cut to the action itself. ~~Start to~~ get to work![d]
try to	I determine to *try to* shovel the walk. You determine to shovel the walk. We all know who gets the chore done.
tend to	~~Tend to~~ avoid *tend to*. Dig straight into the real verb.
in order to	Put on a wool hat ~~in order~~ to keep your ears warm.
point in time	If you say you don't want to shovel *at this point in time*, you sound like you're running for office. Admit that you don't feel like doing it *now*. Remember, though, that the snow will be heavier and crustier if you go after it ~~at some future point in time~~ **later**.
period of time	You have too much to accomplish to waste even the brief ~~period of~~ time it takes to say *period of*.
in light of, in spite of, in terms of (and *of* in general)	~~In spite of its innocent looks,~~ the **innocent-looking** little word *of* ("in anything other than small doses") has been called "among the surest indications of flabby writing."[e] ~~In terms~~

---

d. I owe this tip to Jack Hart, who says, "We all ... share a compulsion to describe the beginnings of actions, rather than the actions themselves ... The beginning of any action is an infinitely brief point. You can visualize the action itself ... but the beginning forms no image at all ... [So rather than write] 'The plane began to circle' ... simply write ... 'the plane circled.'" (Hart, *Storycraft*, 118.)

e. Garner, *Garner's Modern American Usage*, 585–586.

~~of tight writing,~~ strong writers use few *of*s. Stay strong! Avoid the preposition *of*, ~~in spite of its siren song~~! Did you ever consider *of* that big ~~of~~ a deal? Don't worry about *of*, though, when it contributes to a phrasal verb's idiomatic meaning, as in *George didn't know what to make of that day's snowfall.* In this example, *of* acts not as a preposition but as a verb particle.[f] For ~~a whole lot of discussion~~ **more** on verb particles, see "You Don't Know From Prepositions" on page 49.

may or may not	You ~~may or may not~~ **might not** finish the job today.[g]
proverbial	Nothing clutters like a cliché—except the term *proverbial*, which adds clutter to clutter. You can't freshen a stale phrase, like *a breath of fresh air*, by calling it a proverbial breath of fresh air. You'd have better luck clearing away snow by tossing on more snow. Send the whole phrase sailing.
different	If you're of two ~~different~~ minds about whether to use the term *different*, take a break, rest your arms, and mosey over to page 41, "Let Me Count the—Different?—Ways."
weak *be*-verbs	If you haven't taken that break yet, ~~be smart~~ **wise up** and take it now. You need to know about the hazards of *is*, *are*, and other weak verbs, as described on page 13, "*To Be* or Not *To Be*."

---

f. ... unless George (in not knowing what to make of that day's snowfall) is debating whether to make a snowman or an igloo, in which case the verb is neither phrasal nor idiomatic—it's not *make of* but simply *make*—and the *of* acts as a preposition after all.

g. The phrase *may or may not* raises two problems: (1) The *may or* is redundant because the *may not*, all by itself, implies both possibilities. (2) *May not* raises questions of permission. "The phrase should probably be *might not*" (Garner, *Garner's Modern American Usage*, 529). (Wonder why most of this chapter's footnotes have letters instead of numbers? The reason has to do with a software limitation. You get points for noticing.)

false *buts*	A false *but* (a *but* that contradicts nothing) leads people astray like a sign pointing the wrong way. Example: *People could slip on that sidewalk, but I don't feel like clearing it.* This *but* implies a contradiction, ~~but~~ **where** no contradiction exists. A true *but* indicates contradiction: *I wish I felt like clearing that stinking sidewalk, but I don't.*
particular	This ~~particular~~ word only gets in your way. Over the shoulder it goes!
any other words that you can pitch	Going after impediment words like the ones in this list (or in any of a thousand such lists) merely warms you up. After you chuck these words, stretch, bend, twist, shake out your hands, and tackle the real chore: eliminating ~~all manner of overabundant, superfluous, or worn-out words~~ **verbiage**.
never	I'm kidding. Of course you can say *never*. How else can you tell people what words never to use?

This kind of shoveling takes time. But nothing beats the satisfaction of easing someone's way. That satisfaction awaits you. In fact, if you're not busy right now, how about grabbing a shovel and helping a gal out? Come on now, put your back into it.[57] Brian will be home any minute.

---

57. Unlike your back, writing rules are for breaking. If you're writing poetry or lyrics, say, or if you're conveying a tone or a voice, or if you have any other reason for using "forbidden" words, knock yourself out. Not literally. There, I used an *-ly* word. How do you know whether your rule-breaking works? Seek out the kind of readers you'd like to reach, and ask. Have someone read your words to you out loud. My husband does this for me. Sometimes I'm delighted. Sometimes I'm humbled. I never outgrow the need for these reminders that my writing works only when it works for someone other than me.

You Don't Know
From Prepositions

This is the sort of English up with which I will not put.
—Attributed to Winston Churchill

Quick! What kind of word is *from*?

Bet you said, "Ha! Everyone knows it's a preposition. Must be a trap. Better not say preposition."

We all learned it in grade school: *from* is a preposition. When I sat down to draft this chapter, I didn't intend to overturn this teaching. I set out to write a brief notice that, yes, despite what some teachers say, sentences can end with prepositions. I ended up unlearning some "facts"—laboriously, by way of research, confusion, and resistance— and expanding my perspective. I came to see that prepositions are not necessarily prepositions and that easy labels—who knew?—can obscure deeper truths.

I invite you to join me. Set aside, for a moment, what you know. Open yourself to a discussion unlike any other in this book: a deep exploration of one aspect of grammar to which few people will ever give much thought. Consider the outrageous notion that we can't call *from*, by itself, a preposition or anything else. We can't know what *from* is (we can't know from *from*) until we see what it does. Same goes for *over*, *of*, *around*, and the hundred other words that we have always thought of, automatically, as prepositions. In fact, a preposition is a preposition only when it *acts* as a preposition, when it creates a certain kind of relationship between other words in a sentence.

This thing that we've always called a preposition is a verb particle—not a preposition—when it acts as a verb particle, and it's an adverb—not a preposition and not a verb particle—when it acts as an adverb.

In this chapter, I hope to leave you with nothing less than the exhilaration of a new way of seeing language. As a bonus, I wrap up with a few guidelines that enable you—within the admittedly narrow realm of this special group of words—to write with more confidence and freedom.

Preposition, Verb Particle, or Adverb?

Let's start with some definitions based on the way these three word types behave.

- **A preposition** typically appears immediately before—in *pre-position* to—a noun phrase. The preposition connects the noun phrase to another word in the sentence. In *The fox leaped into the river*, the preposition *into* connects *the river* back to *leaped*. The prepositional phrase *into the river* modifies the verb *leaped*. (Incidentally, some linguists no longer even count prepositions among English parts of speech. For grammar lovers, this news ranks up there with the deplanetization of Pluto.[58])

- **A verb particle** combines with a main verb, and sometimes with other particles, to create a multiple-word verb with an idiomatic meaning, a meaning different from that of the individual words. For example, *in* is a verb particle—not a preposition and not an adverb—in *chip in* (help). *Out* is a verb particle in *hand out* (distribute). *Out* and *of* are both verb particles in *drop out of* (quit). *From* is a verb particle in *know from* (understand, have a clue about).

- **An adverb** modifies a verb, an adjective, or another adverb. Adverbs commonly tell when or where or how something happens. In *The couple strolled outside*, the adverb *outside* tells where the couple strolled. Here, *outside* is neither a preposition

---

58. Pluto comparison courtesy of my grammar-loving sister. The reclassification of traditional parts of speech requires a new vocabulary to which no footnote can do justice. For an introduction to this topic, see "A Modern Take (Is *Take* a Noun?) on Parts of Speech" on page 61. See also the bumper sticker "Honk if Pluto is a planet."

(it has no object) nor a verb particle (it contributes to no idiomatic meaning).

The authors of *Analyzing English Grammar* summarize the distinctions this way: "prepositions have noun or noun phrase objects; verb
particles are essential to the meaning of the verb; and adverbs often
can be...deleted."[59]
What kind of word is *up* in each of the following sentences?

> Jack ran up a huge hill.
> Jack ran up a huge bill.
> When he got to the hill, Jack ran up, turned around, and ran back
> down.

Let's zoom in on these. Brackets indicate the verb.

> Jack [ran] **up** a huge hill.

This *up* is a preposition because it connects a noun phrase (*a huge
hill*) to another word in the sentence (*ran*).

> Jack [ran **up**] a huge bill.

This *up* is a verb particle because it is essential to the meaning of the
verb. The words *ran* and *up* function as a unit—as a single verb—with
an idiomatic meaning: "incurred." *Ran* without *up* makes no sense; a
person can't run a huge bill.

> When he got to the hill, Jack [ran] **up**, turned around, and ran
> back down.

59. Thomas P. Klammer, Muriel R. Schulz, and Angela Della Volpe, *Analyzing English Grammar*, 5th ed. (New York: Pearson Longman, 2007), 117–123. I'm grateful to Drs. Klammer, Schulz, and Della Volpe for their feedback on this chapter and on "A Modern Take (Is *Take* a Noun?) on Parts of Speech" on page 61.

This *up* is an adverb because it tells how Jack ran. *Up* has no noun or noun-phrase object, so it can't be a preposition. And *up* is not essential to the meaning of the verb—*ran* still means "ran" without it—so it can't be a verb particle.

Verb Particles and Phrasal Verbs

Verbs that include verb particles—*fight off, come up with, run out of*—are called phrasal verbs because they are phrases. Even when separated, the main verb and its particle function as a unit. In *Jane took the idea in*, the verb is *took in*.

When I first read the term *verb particle*, I pictured a plastic model of an atom, a few colored balls held together by sticks. The main verb (*chip, give, drop*) is the nucleus. One or more verb particles (*in, out, out + of*) are electrons. The whole phrasal verb (*chip in, give out, drop out of*) is the atom.

English includes thousands of phrasal verbs, each with its own idiomatic meaning. (*Cut it out* has nothing to do with using scissors.) Some phrasal verbs have multiple meanings. (*Check out* could mean "look at," "go to a cashier," "exit physically," or "exit mentally." *Put on* could mean "don clothes," "josh a person," "apply makeup," or "play recorded music.") Phrasal verbs can include nouns too; I especially like the oddly perfect *wrap your head around* (understand). The website UsingEnglish.com defines some 2,000 phrasal verbs—a mere sampling—based on 153 main verbs. *Get*, alone, spawns 167 phrasal verbs like these: *get back, get ahead of, get along with, get down, get over.*[60]

> The main verb (*chip, give, drop*) is the nucleus. One or more verb particles (*in, out, out + of*) are electrons. The whole phrasal verb (*chip in, give out, drop out of*) is the atom.

We use phrasal verbs all the time. They give our language color and make it endearingly flexible. The *New York Times* crossword puzzle

---

60. "Phrasal Verb Quizzes—By Verb," *UsingEnglish.com*, last modified June 7, 2012, http://www.usingenglish.com/reference/phrasal-verbs/quizzes-verbs.html.

wouldn't be the *New York Times* crossword puzzle without them. Phrasal verbs also make English maddening to learn. One nonnative speaker calls them "English mutant monsters."[61]

When Is a Word "Essential to the Meaning of the Verb"?

The quirkiness of phrasal verbs can make it tough to tell whether you're looking at a verb particle (essential to the meaning of the verb) or a preposition (not essential to the meaning of the verb), especially when the slippery little word in question precedes a noun. The whole question hangs—hangs, I tell you!—on whether that noun is the object of a preposition or the direct object of a transitive phrasal verb.

Quick review: A transitive verb is a verb that has a direct object; the verb *trans*fers action to a noun (*trans* = "across"). For example, in *The dancer broke in the new shoes*, the phrasal verb *broke in* is a transitive verb, and *shoes* is its direct object (the noun to which it transfers action). Here, *in* is a verb particle. An intransitive verb has no direct object. In *The plate broke into little pieces*, the verb *broke* is intransitive; *pieces* is not a direct object of the verb but an object of the preposition *into*.[62]

In the following examples, the slippery little word *with* must be either a verb particle or a preposition—the only two grammatical possibilities—but which? Does the verb's meaning require *with*, making *with* a verb particle? Or does the verb's meaning not require *with*, making *with* a preposition?

> Go play with those kids.
>
> Don't mess with those kids.
>
> Let's make up with those kids.

The analysis gets tricky.

---

61. Bonnie Trenga, "Phrasal Verbs," *Grammar Girl* blog, July 4, 2008, http://grammar .quickanddirtytips.com/phrasal-verbs.aspx.

62. Some verbs can play either a transitive or an intransitive role. The verb *broke*, for example, is intransitive in *The plate broke into little pieces* (no direct object) and transitive in *The shopper broke the plate* (direct object = *plate*).

Go [play] **with** those kids.

This *with* is not essential to the verb's meaning. Without *with, play*—"frolic"—still pertains to the sentence. So *play with* is not an idiom; it's not a phrasal verb; *with* isn't its particle. *With* is a preposition. *Kids* is an object of the preposition. This sentence has no direct object. The verb, an intransitive verb, is simply *play.*

(*Play with* can be idiomatic. Take *Go play with those ideas. With* is essential to this verb's meaning. Here, *play with* is an idiom meaning "consider." Without *with, play*—"frolic"—no longer pertains to the sentence. So *with* is a verb particle. This sentence has no preposition. *Ideas* is a direct object of the transitive verb *play with.*)

Don't [mess **with**] those kids.

This *with* is essential to the verb's meaning. *Mess with* is an idiom meaning "bother." Without *with, mess*—as in "make messy"—no longer pertains to the sentence. So *with* is a verb particle. This sentence has no preposition. *Kids* is a direct object of the transitive verb *mess with.*

Let's [make **up**] **with** those kids.

Here, the verb particle (*up*) butts up against a preposition (*with*). Each needs its own analysis:

- This *up* is essential to the verb's meaning. *Make up* is an idiom meaning "reconcile." Without *up, make*—"create" or "force"—no longer pertains to the sentence. So *up* is a verb particle. The verb here is *make up.*

- This *with* is not essential to the verb's meaning. Without *with, make up*—"reconcile"—still pertains to the sentence. So *with* is not a verb particle; it's a preposition. *Kids* is an object of the preposition. This sentence has no direct object. The verb is intransitive.

Is this a hoot or what?

The Context Beyond the Sentence
Quick! What kind of word is *along* in this sentence?

> Get along, little dogies.

This *along* can't be a preposition because it has no noun object. (The speaker is neither telling someone to "get along the little dogies" nor telling the little dogies to get along something.) So *along* must be either an adverb or a verb particle. But which?

In fact, we can't say. The sentence doesn't tell us enough. We lack context again, as we did in the beginning (What kind of word is *from*?). We need to know what the speaker means before we can classify *along*.

> [Get] **along**, little dogies.

If the speaker wants the little dogies to stop dawdling, *along* is not essential to the meaning of the verb. You know this because when you delete *along*, the verb *get*—as in "git"—still pertains to the meaning of the sentence. So *along* is an adverb.

> [Get **along**], little dogies.

If the speaker wants the little dogies to stop squabbling, *along* is essential to the meaning of the verb. You know this because when you delete *along*, the verb *get*—as in "git"—no longer pertains to the sentence. *Get along* is idiomatic; it's a phrasal verb meaning "cooperate." So *along* is a verb particle.

A Few Guidelines
Why bother with all this brain-taxing analysis? For me, wrapping my head around this stuff is its own reward. "Getting it" has been a blast. "Getting it" has also brought a few guidelines into focus, clarifying certain decisions. Mostly they're small, but when it comes to writing,

"no decision is too small to be worth wrestling with."[63] Here, then, are those guidelines.

Use Phrasal Verbs on Purpose

When a phrasal verb suggests itself, your first decision is whether to use it. Consider several factors. In their favor, these idiomatic verbs are "natural-sounding" and "lend a relaxed, confident tone," as Bryan Garner, ahem, points out.[64] On the other hand, he continues, they increase word count, so "some rhetoricians prefer avoiding them—hence *handle* instead of *deal with*, *resolve* instead of *work out.*"

Also weighing against phrasal verbs, sometimes, is their informality. Bonnie Trenga, the author of *Off-the-Wall Skits with Phrasal Verbs*, gives this example: "If you were writing a dissertation on Henry VIII, you might not want to write, 'The king hung out with all the nobles.' It would probably be better to write, 'The king associated with all the nobles.' If there's a doubt, use more formal language."[65]

A summary of all this wisdom amounts to a counsel of perfection: choose terms—phrasal or not—that convey your meaning precisely and tightly and that hit exactly the right level of diction for every conceivable audience and purpose.

Do I hear a chorus of angels?

Put Spaces Before Verb Particles

Verb particles and their verbs may cozy up to each other to form compound nouns, but in verb form—that is, in phrase form—they need their space.

> Nouns: *giveaway, hangout, shake-up*
>
> Verbs: *give away, hang out, shake up*

---

63. William Zinsser, *On Writing Well: An Informal Guide to Writing Nonfiction*, 4th ed. (New York: Harper Perennial, 1990), xiii.

64. Garner, *Garner's Modern American Usage*, 628–629.

65. Trenga, "Phrasal Verbs."

Put Spaces After Verb Particles

Quick! *Sylvia went onto the stage* or *Sylvia went on to the stage?*

Have you ever wondered when to put a space between *on* and *to?* Between *in* and *to?* Between *up* and *on?* Wonder no more. If you're looking at a verb particle, put a space after it.

> After entering the movie theater, Sylvia [went] **onto** the stage.

This *on* (the syllable you pronounce) is not essential to the meaning of the verb. You know this because when you delete *on*, the verb *went*— as in "walked"—still pertains to the meaning of the sentence. So *on* is not a verb particle. It's part of the preposition *onto* (no space after *on*).

> After acting in many movies, Sylvia [went **on**] to the stage.

This *on* is essential to the meaning of the verb. You know this because when you delete *on*, the verb *went*—as in "walked"—no longer pertains to the meaning of the sentence. *Went on* is idiomatic; it's a phrasal verb meaning "proceeded." So *on* is a verb particle. It's not part of a preposition. You put a space after it (*on to* not *onto*).

Quick! Does a company's expertise fit into, or fit in to, an online conversation? Hint: Delete *in*. Does *fit* (meaning "be the right size and shape for") pertain? No. *Fit in*, here, is idiomatic. It means "go together or harmonize." Expertise *fits in* to a conversation.

Now do you feel clued into, that is, clued in to, the use of spaces?

Avoid Extraneous Verb Particles

Verb particles sometimes crash the party, sneaking in where they don't belong. Instead of *separating* things *out, separate* them. Rather than *focus in* on something, *focus* on it. Don't *weigh up* your priorities; *weigh* them. When in doubt about a verb, research it out. I mean, research it. Dictionaries cover thousands of phrasal verbs. Some dictionaries cover nothing but phrasal verbs.

Stay Dialed, Dude

Fashionable phrasal verbs often absorb their particles. If you use slang in your writing, despite the risks,[66] stay alert to changes. Once upon a time, we were *bummed out*; these days, we're *bummed*. Back in the day, you'd challenge your rivals to *bring it on*; today, you say *bring it*. Hipsters don't *deal with* problems; they *deal*. They don't get *psyched up*; they get *psyched*. They don't *walk out on* people; they *walk*. When Trailblazer LaMarcus Aldridge "twines eight in a row from twelve feet,"[67] he is not *dialed in*; the man is *dialed*.

End a Sentence with a Preposition If You Need To

The so-called rule against ending a sentence with a preposition has been called a "durable superstition,"[68] a "remnant of Latin grammar,"[69] an "artificial 'rule,'"[70] and "one of the top ten grammar myths."[71] One editor reports having seen many a "tangled sentence due to reluctance to end a sentence with a preposition."[72] That kind of tangled sentence is exactly what some anonymous scribbler back in Churchill's day was railing against when he or she penned the variously cited, often misinterpreted quip ending so magnificently in "up with which I will not put."[73]

---

66. Some style guides advise avoiding slang altogether, but you might want to use it if it suits your audience and purpose. Consider whether the advantages of immediacy and color outweigh the risks of alienating some readers and sacrificing the long-term relevance of your writing. When you nail it, slang, like other "ephemeragy," is "one of the most stimulating devices in the writer's toolbox" (Plotnik, *Spunk & Bite*, 234–235).

67. Example courtesy of my husband, whose encouragement and editorial suggestions have helped shape this whole chapter (not to mention the whole book).

68. Johnson, *The Handbook of Good English*, 386.

69. Garner, *Garner's Modern American Usage*, 654.

70. Paul Brians, *Common Errors in English Usage* (Wilsonville, OR: William, James & Co., 2003), 73.

71. Mignon Fogarty, "Ending a Sentence with a Preposition," *Grammar Girl* blog, March 31, 2011, http://grammar.quickanddirtytips.com/ending-prepositions.aspx.

72. Shauna Roberts, "Phrasal Verbs: Cool, but Often Misused," *Shauna Roberts' For Love of Words* blog, March 26, 2008, http://shaunaroberts.blogspot.com/2008/03/phrasal-verbs-cool-but-often-misused.html.

73. Some researchers question the Churchill attribution. The source usually cited, an anecdote in Sir Ernest Gowers's *Plain Words* (1948), rests on mere hearsay: "*It is said that* Mr. Winston Churchill once made this marginal comment" (emphasis mine). For

I add my voice to the choir in making, finally, the point that I originally set out to make. Feel free to end a sentence with a preposition—but only if you can't find a better word to end it with.

Now You Know From Prepositions—So What?

What does it matter that you now know from prepositions?

Who's going to notice when you confidently put a space between *on* and *to*, or when you freely place *with* wherever you please? Who will appreciate that you've seen through the label "preposition" to deeper linguistic truths? What critic will ever be impressed by such private indicators of your growing mastery over your craft?

> One editor reports having seen many a "tangled sentence due to reluctance to end a sentence with a preposition."

Answer: The only critic with the power to hold you back.

---

more on the debunking of this probably false Churchillism, see Benjamin G. Zimmer, "A Misattribution No Longer To Be Put Up With," *Language Log* blog, December 12, 2004, http://itre.cis.upenn.edu/~myl/languagelog/archives/001715.html. See also Paul Brians, "'Churchill' on Prepositions," accessed May 14, 2012, http://public.wsu.edu/~brians/errors/churchill.html.

A Modern Take (Is Take a Noun?) on Parts of Speech

The adjective is the banana peel of the parts of speech.
—ATTRIBUTED TO CLIFTON FADIMAN

What is a part of speech? You might not believe how much disagreement and nuanced analysis surrounds that question.

This essay ventures into some philosophical questions—what does it mean to classify a word, and how and why have those classifications changed?—before emerging with writerly advice. I find this excursion invigorating, like a deep-sea search for treasure. Come along, and we'll share the spoils.

According to one modern school of linguistic thought, only four word types—nouns, verbs, adjectives, and adverbs—now qualify as parts of speech. Four. The nerve! These—*the*—parts of speech (also called form-class words, as we'll explore deeper down) comprise "the vast majority of words in the English language."[74] The other word types traditionally considered parts of speech—prepositions, pronouns, conjunctions, and interjections (give or take a part)—no longer belong to the club.

Sounds crazy at first. But looking at word types in this new way makes sense, and it solves a problem. The problem with the traditional, more inclusive classification scheme, which remains in common use, is the schizophrenic way it defines word types: "sometimes on meaning,

---

74. Klammer, Schulz, and Della Volpe, *Analyzing English Grammar*, 91. I'm grateful to this book's coauthors for their feedback on this chapter and on "You Don't Know From Prepositions" on page 49.

sometimes on function."[75] Is *take* a noun? The traditional take on parts of speech—the old classification scheme—lacks the distinctions required for answering this question precisely. *Maybe. Sometimes. It depends.* We need a better way to talk about words: a better metalanguage.

The modern take gives us that better metalanguage—one that hinges on new distinctions related to form and function.

What do *form* and *function* mean when it comes to words? All linguists, traditional and modern, agree on this answer.

- **Form:** A word's form is its "physical shape."[76] Form is "what we see or hear when someone uses a word."[77] The form of the word *sidewalk* is s-i-d-e-w-a-l-k. If you add or delete or change the letters—whether meaningfully (*sidewalks*) or randomly (*sidewalkqwerty* or *qwertysidewalk* or *sdqwertywks*)—you change its form.

- **Function:** A word's function is the grammatical role it plays in a phrase or sentence. *Sidewalk* plays one role in *Let's shovel this sidewalk* and another in *I've got the sidewalk blues.*

Notice that we haven't yet asked what kind of word *sidewalk* is; we haven't yet attempted to classify this word. We're setting up a framework for classifying words. (Keep your oxygen flowing. The world we're descending into is esoteric if also wondrous.)

In this crazy language that is English, when you get a hankering to classify a word, you have to look at both its form and its function: the word in itself and the word in the context of other words around it. For example, in *Call me a shoveling fool*, the word *shoveling* is a verb in form (it ends in *-ing*) and an adjective in function (it modifies the noun *fool*). The reality that "form and function do not always match in English"

---

75. Garner, *Garner's Modern American Usage*, 910.
76. Klammer, Schulz, and Della Volpe, *Analyzing English Grammar*, 179.
77. Muriel R. Schulz, e-mail to the author, June 2, 2012.

has been called "the despair of grammar students."[78] We're stuck with it. Coming to terms with the form-function duality—understanding that the question *Is* take *a noun?* has two parts—"is essential for comprehending how English works."[79]

(It's also essential for understanding how the *New York Times* crossword puzzle works. What do I mean? For pointers to examples, look up *crossword puzzle* in the index.)

In a sense, sometimes, you can classify a word by default just by looking at its form in isolation. *Sidewalk*, for instance, can be called a noun in form. How can we say that? This turns out to be a good question. Answering it, if you're a discerning linguist, requires invoking tests for nounness that go beyond the person-place-or-thing definition.

> According to one modern school of linguistic thought, only four word types—nouns, verbs, adjectives, and adverbs—now qualify as parts of speech. Four. The nerve!

Here's one test: *Does adding an* s *create a plural in natural usage?* (Yes, *sidewalks* is a perfectly pedestrian plural. One point for nounness.) A couple of similar tests later, and you have enough clues to determine not only whether a word is a noun in form but how nouny it is. Nounness, it turns out, has a continuum. A noun may be true to form in all ways or in only some ways.

Linguists apply tests to single words (they look for "features of form"[80]) the way chemists apply "a series of chemical tests to identify an unknown substance."[81] They have tests for all four types of form-class words, those newly narrowed parts of speech: nouns, verbs, adjectives, and adverbs. For our purposes, which include not drowning, we need to know only this: the form-class foursome have something in

---

78. Klammer, Schulz, and Della Volpe, *Analyzing English Grammar*, 85.
79. Ibid., 12.
80. Ibid., 91.
81. Ibid., 86.

common that differentiates them from the erstwhile parts of speech. The differentiator? At the risk of oversimplifying—stay away from the derivational and inflectional morphemes!—I'll put it this way: form-class words can change form in predictable ways and still make sense. If you tack on an *s*, an *ing*, or an *est* at the end, or a *pre* or an *ultra* at the beginning, you get words that you'd find in a dictionary. *Sidewalk* becomes *sidewalks. Shovel* becomes *shoveling. Cold* becomes *ultracold.* No one so much as blinks.

The ex–parts of speech? No can do. Take a preposition (*of*), a pro-noun (*it*), or a conjunction (*and*). If you change the form of these words in the usual ways (*ofing, itest, ultraand*), you get something that Merriam-Webster won't go near. In natural usage, these words have one form only.[82]

Because these ex–parts of speech, unlike the now-parts, have only one natural form, it would make no sense to tack on a prefix or suffix to test a word for, say, prepositionness or pronounness. A given word, like *from* or *she*, might usually act as a preposition or as a pronoun, and so we often comfortably (unthinkingly) apply these labels. We jump to classification based on form alone. But in these cases, from a linguist-as-chemist point of view, the word itself gives nothing away. These words have no features of form to test against. Calling a word a preposition in form or pronoun in form would have no meaning. We can't call any word of this type true to form.

What else makes the ex-parts unique? Whereas form-class words (also called content words: *tree, run, fabulous, dizzily*) contain "lexical meaning" in themselves, the ex-parts (*of, it, and*) contribute "grammatical meaning" to a sentence.[83] (What does *of* mean? You see?)

To acknowledge the uniqueness of these ex-parts—of all word types that lack parts-of-speechness—modern linguists give them a club of their own: the structure class. If form-class words are parts of speech, structure-class words are connectors of parts of speech.

---

82. Some pronouns, the troublemakers, prove the exception to this generalization. You didn't expect English to lie down for this analysis, did you?

83. Klammer, Schulz, and Della Volpe, *Analyzing English Grammar*, 95–96.

Structure-class words, or structure words, hold sentences together and enable them to make sense. Structure words may have no features of form, but oh, baby, do they have features of function—so much so that they're also called function words. They are all about what they do. They have only one reason to get out of bed every morning: to create relationships between other words. They are the matchmakers of the sentence

> **Structure-class words have only one reason to get out of bed every morning: to create relationships between other words. They are the matchmakers of the sentence world.**

world. These words put the diagram in *sentence diagram*. Only the most rudimentary of sentences could exist (*See John jump*) without structure words.

For the record, structure words include prepositions (*with*), pronouns (*he*), conjunctions (*but*), determiners (*the*), auxiliaries (*might*), qualifiers (*very*), relatives (*whose*), and interrogatives (*where*).

From is a structure word in this sentence: *Call me a shoveling fool from Liverpool.* It does what only a preposition can do: it creates a relationship between its object (*Liverpool*) and another noun (*fool*).[84] Without structure-class words, like *from*, to bring coherence to the jumbo jumble of form-class words—that plenitudinous stockpile of disconnected speech parts—you and I couldn't be communing right now because books would not exist. Isn't it time that words this uniquely and powerfully endowed had a class of their own?

Compared with the monocular traditional view of parts of speech, the binocular modern take, with its distinction between form-class and structure-class words, gives us a more accurate and useful way to communicate, and think, about these multidimensional critters known as words. The new metalanguage provides a more meaningful peek into the way language works.

---

84. For more on *from* and its fellows, including more reasons that we can't automatically call *from* or any other word a preposition, see "You Don't Know From Prepositions" on page 49.

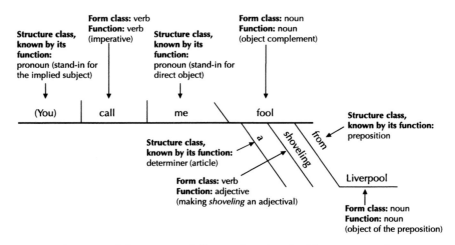

Diagram of the sentence Call me a shoveling fool from Liverpool.

What's more—here we emerge from the sea with gold in hand—this distinction between form-class and structure-class words gives writers new insight into an ancient tool for enlivening sentences, a tool that yields "instances of wonderfully imaginative language use."[85] The tool goes by various names, including *anthimeria* (with or without the *h*), *grammatical shift*, and *enallage*. These terms all get at the same thing: the twisting of a word's expected usage—the yanking of words out of what linguists call their natural classes.

You can accomplish this twisty yankiness in two ways:

- Add a prefix or suffix, thereby changing the word's form (*parts-of-speechness*).

- Plug the word into a mind-pricklingly surprising position in a sentence, thereby changing its function (*On the count of three, everybody enallage*).

---

85. Klammer, Schulz, and Della Volpe, *Analyzing English Grammar*, 89.

Arthur Plotnik, who writes lively books on lively writing, gives this playful example: "So how can questy writers enallage their way to the big Fresh? On that little how, we could noun and verb you all day."[86]

How does the distinction between form-class and structure-class words ratchet up our enallagitudinous abilities? It points us straight to the pay-off words: the form-class words. Words of these types (*doughnut, bristle, cheeky, abundantly*), with their continua of form, afford the most play. A glance at Lewis Carroll's famous poem "Jabberwocky" ("All mimsy were the borogoves, And the mome raths outgrabe") snapdoodles this point home.

> **When you need an especially powerful word, when you want to blast a word beyond Merriam-Webster's reach, enallage from the form classes. Brilliantificate nouns, verbs, adjectives, and adverbs.**

When you need an especially powerful word, when you want to blast a word beyond Merriam-Webster's reach, enallage from the form classes. Brilliantificate nouns, verbs, adjectives, and adverbs. Don't waste your time on structure words.

Unless you're itching to say something like, "From me no froms." We are, after all, talking English here.

I can't hope to answer all the questions you may be asking yourself right now, like *What is an inflectional morpheme anyhow?* or *How do you diagram sentences?*[87] Mainly, I want to pull this one message out of the bottle: when it comes to parts of speech, you might not know what you think you know. As a writer in search of new life for old words, you just might decide to consider that discovery good news.

---

86. Plotnik, *Spunk & Bite*, 113.

87. To find out about inflectional morphemes, the distinction between form classes and structure classes, and the intricacies of diagramming sentences, see *Analyzing English Grammar*. For insights specifically into diagramming ("unscrewing") sentences, see Kitty Burns Florey, "Taming Sentences," June 18, 2012, *New York Times* online series *Draft*, http://opinionator.blogs.nytimes.com/2012/06/18/taming-sentences, or Kitty Burns Florey, *Sister Bernadette's Barking Dog: The Quirky History and Lost Art of Diagramming Sentences* (Orlando: Harcourt, 2006).

PART II

Up with Sentences and Paragraphs

A sentence should contain no unnecessary words, a paragraph no unnecessary sentences, for the same reason that a drawing should have no unnecessary lines and a machine no unnecessary parts.
—STRUNK AND WHITE, *ELEMENTS OF STYLE*

The Last Word

If you want a happy ending, that depends, of course, on where you stop your story.
—ORSON WELLES IN *ORSON WELLES INTERVIEWS*

Plants are fueled by a simple sugar that results from a magic combination of sunlight, water, and carbon dioxide: glucose. To borrow the title of a Dylan Thomas poem, glucose is "the force that through the green fuse drives the flower." When this sweet power source runs low, a plant experiences a chlorophyll shortage, which triggers it to do something remarkable. A plant running on empty pours most of its energy into the latent buds at its branch tips. Gardeners call this phenomenon *terminal dominance.*

Did you notice that, like sugar-deprived plants, all the sentences in the previous paragraph push their energy to the terminus? In sentence after sentence, the most important word appears just before the period. All these sentences exemplify a powerful principle that writers have followed for centuries, a principle that the iconic pair William Strunk Jr. and E. B. White (whose thoroughly highlighted, falling-apart 1972 edition of *The Elements of Style* taught me much of what I know about powerful writing) sum up as follows: "Place the emphatic words of a sentence at the end."[88]

Bryan Garner, who agrees that "the punch word in a sentence should come at the end" urges writers to read their sentences aloud, "exaggerating the last word in each sentence. If the reading sounds awkward or

---

88. William Strunk Jr. and E. B. White, *The Elements of Style*, 2nd ed. (New York: MacMillan, 1972), 26. Note: In the 2005 3rd edition, whimsically illustrated by Maira Kalman, this statement falls on page 52.

foolish, or if it seems to trail off and end on a trivial note, the sentence should probably be recast."[89]

Note the strategic location of Garner's operative word, *recast.*

Emphatic endings can produce periodic sentences, which are long-ish, usually complex constructions that gradually, after the unfolding of a series of delaying phrases, not unlike the ones you're reading right now, wind their way toward the main point. Writers from Cicero to Henry Wadsworth Longfellow to Barack Obama have created masterful periodic sentences in every genre, from oratory to poetry to memoir, always bringing the reader to a full stop on a word that resonates.

> A plant running on empty pours most of its energy into the latent buds at its branch tips. Gardeners call this phenomenon *terminal dominance.*

For example, the night he won the 2008 US presidential election, Barack Obama did not open his speech in Chicago by saying, "Tonight is your answer, in case you wondered if the dream of our founders is still alive in our time." He opened with a periodic build: "If there is anyone out there who still doubts that America is a place where all things are possible, who still wonders if the dream of our founders is alive in our time, who still questions the power of our democracy, tonight is your answer."[90]

I recommend mastering the periodic sentence yourself. Have fun with it, even. Beware, though. If you overuse this structure, strewing clause after clause (whether independent or dependent) in front of your main ideas, front-loading too many sentences in a row, withholding gratification over and over, your readers (who, if you're lucky, want to find out what you're getting at) may stray from what you're saying and focus, as you may be doing right now, on the syntax.

---

89. Garner, *Garner's Modern American Usage*, 734.

90. "Transcript: 'This Is Your Victory,' Says Obama," CNN Politics, November 4, 2008, http://articles.cnn.com/2008-11-04/politics/obama.transcript_1_transcript-answer-sasha-and-malia?_s=PM:POLITICS.

For example, unless your name is Nikolai Gogol, you probably want to avoid giving any sentence this much of a wind-up:

> Even at the hour when the grey St. Petersburg sky had quite dispersed, and all the official world had eaten or dined, each as he could, in accordance with the salary he received and his own fancy; when all were resting from the departmental jar of pens, running to and fro from their own and other people's indispensable occupations, and from all the work that an uneasy man makes willingly for himself, rather than what is necessary; when officials hasten to dedicate to pleasure the time which is left to them, one bolder than the rest going to the theatre; another, into the street looking under all the bonnets; another wasting his evening in compliments to some pretty girl, the star of a small official circle; another—and this is the common case of all—visiting his comrades on the fourth or third floor, in two small rooms with an ante-room or kitchen, and some pretensions to fashion, such as a lamp or some other trifle which has cost many a sacrifice of dinner or pleasure trip; in a word, at the hour when all officials disperse among the contracted quarters of their friends, to play whist, as they sip their tea from glasses with a kopek's worth of sugar, smoke long pipes, relate at times some bits of gossip which a Russian man can never, under any circumstances, refrain from, and, when there is nothing else to talk of, repeat eternal anecdotes about the commandant to whom they had sent word that the tails of the horses on the Falconet Monument had been cut off, when all strive to divert themselves, Akakiy Akakievitch indulged in no kind of diversion.[91]

Read that sentence in one breath!

By the way, the term *periodic* doesn't always imply a period. Any independent clause (that is, any complete thought that stands alone grammatically) can be periodic, even if it ends with a colon, a semicolon, or a dash. If a period *could* go there, so can the word with the most punch; you don't have to come to a literal point to come to your point.

---

91. Nikolai Gogol, "The Overcoat" (1842), accessed May 11, 2012, http://www.eastoftheweb.com/short-stories/UBooks/Over.shtml.

Periodic structure applies also to multisentence paragraphs. In a periodic paragraph, the topic *sentence* comes last. Take the following periodic paragraph from Jack Hart's book *Storycraft*. Hart builds anticipation sentence by sentence so that the final words—words that reflect a critical journalistic decision—detonate.

> But what product [should the reporter focus on? He] considered several and finally landed on a doozy. It flowed to the Pacific Rim in vast quantities, where sales depended on the growing middle class, the group most threatened by the economic crash. Our region dominated its production. It was familiar, low-tech, non-threatening, and easy to understand. If anything linked the Pacific Northwest to the Asian economic crisis, [the reporter] said, it was the McDonald's frozen french fry.[92]

William Zinsser echoes the importance of a paragraph's final sentence, likening it to "the periodic 'snapper' in the routine of a stand-up comic."[93] He says, "These 'snappers' at the end of a paragraph propel readers into the next paragraph."[94]

You can propel your own readers along by treating all kinds of endings—paragraphs, sentences, stanzas, sections, verses, or chapters—strategically. As Strunk and White put it, "The principle that the proper place for what is to be made most prominent is the end applies equally to the words of a sentence, to the sentences of a paragraph, and to the paragraphs of a composition."[95] (Strunk and White missed an opportunity here in concluding their sentence—and in fact their whole chapter—with *composition*. They could have reinforced their the point with positioning: "To strengthen your sentences, paragraphs, and entire compositions, put your prominent information at the end.")

---

92. Hart, *Storycraft*, 184.
93. Zinsser, *On Writing Well*, 66.
94. Ibid., 257.
95. Strunk and White, *Elements of Style*, 2nd ed., 27 (page 53 in the illustrated 3rd edition).

When you're ready to make the Big Point, when you've come to the main thing that you want people to remember or the action you want them to take, when you've reached the climax of your argument, when your most potent word has worked its way down to your fingertips and is practically bursting, you may debate whether to bold it, you may consider all-capping it, you may toy with italicizing it, but you know beyond all doubt exactly where to put it: here.

A Definition Is Where
You Don't Say Is Where

If a train station is where the train stops and a bus station is where the bus stops, what is a work station?
—Unknown

A definition is where you say what something means.

Huh?

Is where. What an unuseful, unsatisfying phrase. I'm talking about definitions like these:

A gravy train *is where* someone makes lots of money without doing much for it.

A retweet *is where* you forward a tweet.

Horticulture *is where* people grow plants.

If you wanted to understand gravy trains, retweets, and horticulture, these half-clarifications would leave you wondering what *types of things* these things are. Surely a gravy train has nothing to do with gravy or trains…does it? What exactly is a retweet before it gets forwarded? If horticulture is where people grow plants, does that make it a plot of ground? You might reread the definitions, suspecting that you missed something. You might feel vaguely cheated, left behind, with the most basic of questions unanswered.

Is where. What a great big abandoner.

I say, abandon it right back. Here's how. Whenever you're struggling to define something, think, *Noun is noun.*[96] The first noun is the thing

---

96. I owe the phrase *Noun is noun* to Penny Jessie Beebe and Sharon Ahlers. During their many years of teaching Engineering Communications at Cornell University,

you want to define. The second noun is the category that the thing belongs to: the type of thing the thing is.

> A gravy train is *a source of income* that ...
>
> A retweet is *a Twitter message* that ...
>
> Horticulture is *the science of ...*

From there, you can go long or short with your definition, depending on what you think your readers will wonder about. You might explain what the thing does, how it works, what parts it comprises. You might include images, analyze word origins, give examples.[97] Or you might not. But don't skip

> **Whenever you're struggling to define something, think,** *Noun is noun.*

over that categorizing noun. A definition is a statement that tells what type of thing a thing is. At least that is where, yes *is where*, a definition starts.

---

they have passed along this tip to hundreds of students—and to at least one fellow teacher, who has now passed it along to you. After reading this chapter, Penny e-mailed (April 7, 2012) to say, "The X is Y definition is one of the most useful tools I know for pressing students to pin down what they're trying to explain. The companion to 'noun is noun,' of course, is 'to verb is to verb.' (When students need a definition of 'verb,' my version is 'anything you can do, visible or invisible,' and a noun is any *thing*, abstract or concrete.) 'Adjective is adjective' doesn't work nearly as well because of the various suffixes, but I use it."

97. For definition examples of all kinds, see Mike Markel, "Drafting and Revising Definitions and Descriptions" in *Technical Communication*, 6th ed. (Boston: Bedford/ St. Martin's, 2001), 219–249.

Metaphors Are Jewels

The greatest thing by far is to have a command of metaphor.
This alone cannot be imparted by another; it is the mark of genius,
for to make good metaphors implies an eye for resemblances.
  —ARISTOTLE, POETICS

Want to wake up your readers? Poke them with a good metaphor. (I just poked you with one. Are you awake?)

Metaphors are handy rhetorical devices that compare one thing to another. Metaphors, also known as comparative tropes, "heighten the meaning or clarity of a subject by relating it to something more vivid."[98] Throw in *like* or *as*—for example, *A metaphor is like a pointy object that you can poke readers with*—and you have a simile, a type of metaphor that makes the comparison explicit. With or without *like*, metaphors help new concepts click and help old ones perk up.

To write is to use metaphors.

To write is, too often, to use metaphors badly.

Metaphors easily go wrong. Consider this egregious example (from *Bad Metaphors from Stupid Student Essays*): "His thoughts tumbled in his head, making and breaking alliances like underpants in a dryer without Cling Free."[99] I'd call this a case of "too much image."[100] Unless you're going for a comic effect, choose comparisons that fit the context.

---

98. Plotnik, *Spunk & Bite*, 63. He goes on here to say, "When the comparison also tickles the reader's fancy, you've got a winner." His whole chapter "The Punchy Trope" (61–70) tickles me right in the fancy.

99. "Bad Metaphors from Stupid Student Essays," *MIStupid.com: The Online Knowledge Magazine*, accessed June 24, 2012, http://mistupid.com/people/page027.htm.

100. Plotnik, *Spunk & Bite*, 34.

Good metaphors enhance; bad ones distract. Boy, do those tumbling underpants distract.

May I have your attention back? I have more bad metaphors for you. It's almost unfair to pick on business speak; it's too easy. I can't resist giving you this example, though. I once saw a conference presentation that started with these three bullets:

- rabbit hole
- brass tacks
- recipes

This was the whole slide. It could have been Slide 1 in any presentation on any topic in any field. As everyone in the room knew, *rabbit hole* implied that the speaker would jump into a complicated topic full of confusion, like Alice's Wonderland. *Brass tacks* implied that the speaker would get down to some basics. *Recipes* implied that the speaker would step through some recommendations.

I could have gone along with any one of those cliché-metaphors—the rabbit hole, the brass tacks, the recipes—by itself without much trouble. (Such metaphors are called dead or dormant metaphors because the images, through overuse, generally go unnoticed.) But yoked together like that, these three metaphors combine to form a bizarre scenario: a bunny sliding down a hole, then pricking itself on some brass tacks, then abruptly concerning itself with cooking. Whoa! As I sat there in that roomful of people, wanting to get my money's worth out of every minute, I was so taken by this extra-mixed metaphor, that I barely processed what the speaker had to say.

Here's another example of business-speak metaphors gone wrong. "Patents [yes, patents] attempt to create a level playing field, but the last thing an 800-pound gorilla of a company wants is a fair fight."[101] Huh? I have no idea what this jumble of images says about patents—or

---

101. R. L. G. (Robert Lane Greene), "Metaphors: Who Wants a Fair Fight with an 800-Pound Gorilla?" *The Economist, Johnson* blog, July 27, 2011, http://www.economist .com/blogs/johnson/2011/07/metaphors.

about playing fields or gorillas. Even if I cared about patents, this clash of clichés would knock the caring right out of me.

A reader sinks into a *good* metaphor comfortably, like Goldilocks easing into Baby Bear's chair. Arthur Plotnik, a writer's writer among writer's writers, describes good comparative tropes (good metaphors) as "factory-fresh, unpredictable, economical, and custom-fitted."[102]

The ideal metaphor works without seeming to do any work at all.

Where can you find examples of good metaphors? Look in plays ("All the world's a stage") and in poems, novels, e-mails, blog entries, brochures, technical manuals, or any other vehicles (!) of human communication. In an article on information architecture, two IBM pros compare the abstract term *information model* to another model that their readers find familiar: a model home. Andrea Ames and Alyson Riley explain, "If you've ever looked at a model home while house-hunting" (yes, I can see that model home now—the gleaming countertops, the fresh flowers) "you know that the model is an example, a pattern that shows an ideal state...Let's take a look at a handful of 'model homes' from the world of information."[103]

The natural, comfortable way that Ames and Riley compare information models to model homes gives me confidence in their knowledge and in their ability to explain what they know. Their introductory metaphor (which goes on for many sentences, making it an extended metaphor) prepares me to understand even as it motivates me to keep reading.

Another effective extended metaphor comes from Carmen Hill, a content strategist for a business-to-business (B2B) marketing agency. In her blog post "Thrillers, Fillers and Spillers: Cultivating Your B2B Content Garden," Hill compares three types of plants needed for an attractive container arrangement (thrillers, fillers, and spillers) to three types of content needed for an attractive B2B marketing program:

---

102. Plotnik, Spunk & Bite, 63.

103. Andrea Ames and Alyson Riley, "Helping Us Think: The Role of Abstract, Conceptual Models in Strategic Information Architecture," *Intercom: The Magazine of the Society for Technical Communication*, January 2012 (volume 59, issue 1), 16–17.

- A thriller—whether in plant or paragraph form—is "a focal point, something big and bold with enough sex appeal to catch the eye and inspire further discovery."
- Fillers are "essential elements that build on the featured attraction."
- A spiller is "anything that can be easily snipped and shared," whether a vine or a video.[104]

Hill develops her comparison smoothly and artfully, choosing broad terms—*focal point, big, bold, essential elements, snipped, shared*—that stretch to fit both plant life and marketing content.

You can find good metaphors in scriptwriting too. In the TV show *House, M.D.*, the main character, Dr. Gregory House, frequently calls on his powers of comparison to explain medical concepts to his team and, conveniently, to the show's viewers. Here, he compares diagnosis to fishing:

> Dr. House: You know, when the Inuit go fishing, they don't look for fish.
>
> Dr. Wilson: Why, Dr. House?
>
> Dr. House: They look for the blue heron, because there's no way to see the fish. But if there's fish, there's gonna be birds fishing. Now, if [the patient]'s got hairy-cell, what else are we gonna see circling overhead?[105]

If only real doctors had scriptwriters like that.

Open any work of literature, and you'll find comparative tropes galore. Tropemaster Mary Karr's first memoir, *The Liars' Club*, serves up delicious metaphors on every page.

---

104. Carmen Hill, "Thrillers, Fillers and Spillers: Cultivating Your B2B Content Garden," *Babcock & Jenkins* blog, July 18, 2011, http://www.bnj.com/thrillers -fillers-spillers-cultivating-b2b-content-garden.

105. Richard Nordquist, "'House' Calls: The Metaphors of Dr. Gregory House," *About.com Grammar & Composition*, 2012, http://grammar.about.com/od/rhetoricstyle/a/house metaphors.htm.

Liars' Club Example 1:

> [Grandma had] started auctioning Mother off to various husbands when she was only fifteen. Like some prize cow...fattened for the highest bidder.[106]

Mother = prize cow. *Prize cow!* Two little words, and we're there. Instantly, we *get* Grandma, and we have no choice but to despise her. Mother's resentment washes over us. With this compact comparison, Karr achieves what many writers live for: she places us—smack—into her characters' psyches.

Liars' Club Example 2:

> The fact that my house was Not Right metastasized into the notion that I myself was somehow Not Right.[107]

Not-Rightness = cancer. This metaphor sneaks in under the cloak of the verb *metastasized*. (Did you notice that in the previous sentence, my own metaphor sneaks in under the cloak of the verb *sneaks*? You can practically see the metaphor

> **My own metaphor sneaks in under the cloak of the verb *sneaks*.**

tip-toeing. Metaphor as cloak-clad agent of surreptitious activity. That's a metaphor for *metaphor*. A *meta*metaphor. Dizzy yet?)

Liars' Club Example 3:

> If Daddy's past was more intricate to me than my own present, Mother's was as blank as the West Texas desert she came from.[108]

Blankness of Mother's past = blankness of West Texas desert. This apt, revealing comparison feels natural, inevitable. If you were reading it in context, you'd whiz right by it. But let's slow down and look at this sentence. Karr could have made any number of comparisons:

106. Mary Karr, *The Liars' Club* (New York: Penguin, 1995), 10.
107. Ibid., 12.
108. Ibid., 23.

"as blank as a whiteboard with nothing written on it" or "as blank as a clueless person's face" or "as blank as a cloudless sky." Lots of things can be blank. She chose a comparison that multitasks in a big way, accomplishing all these things:

- It answers the question, how blank did Mother's past seem? (As blank as a desert.)

- It locates us in Mother's world. (Mother came from the West Texas desert.)

- It conveys that world's emptiness. (The West Texas desert is blank.)

- It communicates the prominence of this desert in the daughter's mind. (She chose it for her comparison, after all. Each metaphor reveals something about the writer or character who chooses it.)

- It reinforces the daughter's longing for closeness with her mother. (A blank desert conjures aloneness, thirst.)

- It serves as a transition. (The first half of the sentence reaches back to the previous paragraphs about the father. The second half gracefully swivels toward the ensuing paragraphs about the mother.)

- It motivates us to keep reading. (Blank West Texas desert. What was it like for Mother to grow up there, and why didn't she talk to her daughter about it? *I must know.*)

We get all that from a single sentence comparing a person's past to a desert—a rich equating of two unequal things. Metaphor. Choose a good one, and you bestow on your readers a shiny nugget of compressed communication.

Go metaphor hunting in any good writer's work. Discover your own gems. Admire them from all sides. Feel their edges. Study the way they gleam. Then don't be surprised when your own writing shows a new sparkle.

Lend Your Commas
a Hand—or Two

Being unemployed and looking through the want ads today, I found it very disconcerting that commas are frequently misused. How, is one, supposed to get, enthusiastic about, applying for, a job, in which the, employer, doesn't even, understand, how to use, commas?
— Attributed to Janis Williams

Next time you wonder whether to use one comma or two to set off a word or phrase in the middle of a sentence, imagine reaching in and lifting that word or phrase out with both hands. Does the sentence still make sense? If so, lower the text back in, and put commas in place of your hands.

For example, you need both commas in all these sentences:

Fruit flies, for example, can breed up to ten times an hour.

The TV, however, sat idle.

My pal Gerry, a baker, learned to love the sunrise.

The house that Sandee likes, the one with the striped curtains and the funny gargoyle on the second story, went up for sale last week.

The phrase *a baker* here is an example of the rhetorical device known as an appositive, a noun or noun phrase that renames the noun directly preceding it. If you can take the appositive out of the sentence and still know which noun you're talking about (we don't need *the baker* to know who Gerry is), grammarians call the appositive nonrestrictive. That means, enclose it in commas.

85

(Words or phrases dropped into the middle of a sentence are called parenthetical—sometimes writers drop in whole sentences parenthetically—after the Greek word *parentithenai,* "to insert." Parenthesis, the insertion of an aside, is a rhetorical device that enables writers to stuff a bit of information into a sentence [using commas, parentheses, dashes, or even square brackets] to provide context or clarification exactly where the reader needs it. Wise writers would never stuff in a whole parenthetical paragraph. Or would they?)

> **Next time you wonder whether to use one comma or two to set off a word or phrase in the middle of a sentence, imagine reaching in and lifting that word or phrase out with both hands.**

With certain types of words, the second comma goes missing especially often. For example, even though most style guides would call for commas on each side of the following parenthetical words (right where you'd put your hands), many writers would omit these second commas.

> The plane will land in Portland, Maine, right on time.
>
> The letter dated January 2, 1987, changed George's life.

In some cases, as in the examples below, you can leave the comma pair out altogether, and some style guides, including *The Chicago Manual of Style*, now recommend doing exactly that.[109] You can use the commas if you like, though—as long as you use them both.

> Macy's, Inc., made headlines today.
>
> *or*
>
> Macy's Inc. made headlines today.
>
> *not*
>
> Macy's, Inc. made headlines today.

---

109. *The Chicago Manual of Style,* 322.

Rodney, Jr., has a birthday coming up.

or

Rodney Jr. has a birthday coming up.

not

Rodney, Jr. has a birthday coming up.

Leaving out half of a pair of commas is like leaving out half of a pair of parentheses. You wouldn't do that (would you?

Running On about Run-Ons

A comma splice walks into a bar, it has a drink and then leaves.
—ERIC K. AULD, *SEVEN BAR JOKES INVOLVING GRAMMAR AND PUNCTUATION*

Teachers: *To use this chapter as an exercise, hand out the* befores.
Writers: *When the* befores *get tiresome, skip them.*

Before: I didn't plan to write about run-on sentences, a much-hyped book by a respected author shocked me into it, run-on sentences, the kind formed by comma splices, litter the pages, it ain't pretty.

After: I didn't plan to write about run-on sentences[.] A much-hyped book by a respected author shocked me into it[.] Run-on sentences, the kind formed by comma splices, litter the pages[;] it ain't pretty.

Before: Think punctuation doesn't affect your bottom line? Ask Amazon, Leo Frishberg, a user-experience pro says that Amazon tests its web pages to compare the effects of sentence-level differences a semicolon vs. a period for example on usability Amazon not only cares about getting the punctuation correct but also measures (read invests money to determine) which of various correct possibilities elicits a better response, powerful testimony!

After: Think punctuation doesn't affect your bottom line? Ask Amazon[.] Leo Frishberg, a user-experience pro[,] says that Amazon tests its web pages to compare the effects of sentence-level differences[—]a semicolon vs. a period[,] for example[—]on usability[.][110] Amazon not

---

110. Paraphrased from Leo Frishberg, "Pillars of the Community: Technical Communication in Service of User Experience Architecture," (lecture, Willamette Valley Society for Technical Communication Chapter meeting, Portland, OR, May 24, 2012).

only cares about getting the punctuation correct but also measures (read[,] invests money to determine) which of various correct possibilities elicits a better response[.] Powerful testimony!

Before: A comma splice is a comma between independent clauses, the comma joins (splices) the clauses together into a run-on sentence, comma splices work well for connecting short independent clauses, as in *I came, I saw, I hopped the bus home*, otherwise, comma splices make reading a chore.

After: A comma splice is a comma between independent clauses[.] The comma joins (splices) the clauses together into a run-on sentence[.] Comma splices work well for connecting short independent clauses, as in *I came, I saw, I hopped the bus home*[.] Otherwise, comma splices make reading a chore.

Before: Some run-on perpetrators dispense with the comma splice altogether they fuse independent clauses together without a wisp of punctuation to give readers a clue one clause merges into the next buffered by nothing but a space some writers use this kind of run-on sentence (a fused sentence) to advantage if you know what you're doing you can too but unless you're after a breathless quality or a stream-of-consciousness effect don't fuse your sentences punctuate them.

After: Some run-on perpetrators dispense with the comma splice altogether[.] They fuse independent clauses together without a wisp of punctuation to give readers a clue[;] one clause merges into the next[,] buffered by nothing but a space[.] Some writers use this kind of run-on sentence (a fused sentence) to advantage[.] If you know what you're doing[,] you can too[.] But unless you're after a breathless quality or a stream-of-consciousness effect[,] don't fuse your sentences[—] punctuate them.

Before: In short, avoid joining independent clauses with either commas or spaces, this is an independent clause this is too each independent clause has a subject and a verb and forms a complete thought.

After: In short, avoid joining independent clauses with either commas or spaces[.] This is an independent clause[.] This is too[.] Each independent clause has a subject and a verb and forms a complete thought.

Before: Luckily, you can fix a run-on sentence, whether it's spliced or fused, in lots of ways. You can always insert a Superman of a separator (a period, a semicolon, a colon, or a dash depending on the relationship between the two independent clauses) you don't have to limit yourself to punctuation marks alone. Get a little wild! Slip in an occasional coordinating conjunction (like *but* or *and*) coupled with a comma you'll win readers' hearts every time.

After: Luckily, you can fix a run-on sentence, whether it's spliced or fused, in lots of ways. You can always insert a Superman of a separator (a period, a semicolon, a colon, or a dash depending on the relationship between the two independent clauses)[, **but**] you don't have to limit yourself to punctuation marks alone. Get a little wild! Slip in an occasional coordinating conjunction (like *but* or *and*) coupled with a comma[, **and**] you'll win readers' hearts every time.

Before: You can also fix run-on sentences by subordinating one of the clauses. In other words, use a subordinating conjunction (like *because*) to turn one of the independent clauses into a dependent clause, exposing a logical relationship. Of course, you can't use this technique to fix every run-on sentence that kind of logical relationship doesn't always exist.

After: You can also fix run-on sentences by subordinating one of the clauses. In other words, use a subordinating conjunction (like *because*) to turn one of the independent clauses into a dependent clause, exposing a logical relationship. Of course, you can't use this technique to fix every run-on sentence [**because**] that kind of logical relationship doesn't always exist.

Before: You may wonder how to determine the best fix for a run-on sentence consider two things the way the parts of the sentence relate to each other and the "tone and rhythm" that fit your context you may want to rewrite the whole thing.[111]

After: You may wonder how to determine the best fix for a run-on sentence[.] Consider two things[:] the way the parts of the sentence relate to each other and the "tone and rhythm" that fit your context[.] You may want to rewrite the whole thing.

Before: If you struggle with run-on sentences, and even if you don't, I recommend Lynne Truss's delectable *Eats, Shoots & Leaves*, which includes joyful explorations of raucously conflicting expert opinions on such prickly subjects as the future of the semicolon as the author herself says, "You know those self-help books that give you permission to love yourself? This one gives you permission to love punctuation."[112]

> I could run on and on about punctuation a person could find worse things to love.

After: Read *Eats, Shoots & Leaves*.

I could run on and on about punctuation a person could find worse things to love.

---

111. Mignon Fogarty, "What Are Run-On Sentences?" *Grammar Girl* blog, August 26, 2010, http://grammar.quickanddirtytips.com/run-on-sentences.aspx.

112. Lynne Truss, *Eats, Shoots & Leaves: The Zero Tolerance Approach to Punctuation*, illustr. Pat Byrnes (New York: Gotham Books, 2008, based on the 2003 British edition), 40.

Touching Words

Words have to be crafted, not sprayed. They need to be fitted together with infinite care.
—ATTRIBUTED TO NORMAN COUSINS

Sentences remind me of those little handheld puzzles with numbered tiles that you slide around. You can solve many sentence problems simply by sliding words around until those that belong together are touching. As Strunk and White say, "Keep related words together."[113] In particular, look for ways to bump related words into each other at these meeting places:

- commas
- colons
- periods, semicolons, or dashes (where independent clauses meet)
- verbs

In the examples below, notice how much more smoothly the sentences read when the first highlighted word or phrase glides into the second.

Meet Me at the Comma

Commas provide a common meeting place for words that belong together. Consider this gem of a botched sentence (found in certain pre-gender-neutralized editions of Strunk and White's *Elements of Style*):

---

113. Strunk and White, *Elements of Style*, 2nd ed., 22 (page 44 in the illustrated 3rd edition).

Before: As a **mother of five**, with another on the way, **my ironing board** is always up.[114]

This sentence cracks me up. Grammatically, if illogically, *mother of five* modifies the far-away subject of this sentence: *my ironing board*. This pairing calls to mind a pregnant ironing board with five little ironing boards running around it. As a first step toward fixing this sentence, scooch *mother* and *board* (modifier and modified) together.

Interim: As an expectant **mother of five, my ironing board** is always up.

Bringing these words together reveals the problem: *mother of five* is a dangling modifier, that is, a word or phrase intended to modify a word that's missing. The word that's missing (we can't help but see now) is the sentence's true subject: *I*.

After: As an expectant **mother of five, I** always have my ironing board up.

In this corrected (if still laughable) sentence, it's no accident that modifier and modified meet at a comma. If you want to win the word-order game, use the comma as a meeting place. Think of it as a curved version of that slim space between numbered tiles.

Word-matching at a comma also fixes the following misplaced modifiers (*nature lover* in the first example and *wearing my pajamas* in the second).

Before: As a **nature lover**, I'm sure **you** would agree that this land is worth preserving. (Here, I'm the nature lover.)

After: As a **nature lover, you** would surely agree that this land is worth preserving. (Here, you're the nature lover.)

---

114. Ibid., 9. (This example, alas, no longer appears in the illustrated 3rd edition.)

> *Before:* One morning I shot an elephant **wearing my pajamas**. (Groucho Marx)
>
> After: One morning, while **wearing my pajamas, I** shot an elephant.

Bringing the right words together at a comma fixes not only dangling and misplaced modifiers but also remote relatives. I love this term. Like a kid who lives across the country from Grandma, a remote relative in grammar is a relative pronoun—*that, which, who, whose*—that "lives" too far away from its antecedent. The following example unites the remote relative, *whose*, with its antecedent, *house*. Remote no more!

> *Before:* Demolition crews have finally removed an old **house** near the Capitol Building, **whose** owners had previously refused to leave. (All US citizens had refused to leave the Capitol Building?)
>
> *After:* Near the Capitol Building, demolition crews have finally removed an old **house, whose** owners had previously refused to leave.

Meet Me at the Colon

Commas aren't the only sites for reunions. Proximity breeds readability at another punctuation mark: the colon. This double-dot symbol—this skinny equal sign—works best when it serves, in fact, as an equal sign, a meeting place for equivalent items. The word touching the colon on the left ideally matches whatever touches it on the right.

Take the following two sentences. The only difference between them is word position.

> *Before:* Sal skimmed these **bestsellers** while at the library: **Unbroken, Freedom, and Words Fail Me**.
>
> *After:* While at the library, Sal skimmed these **bestsellers: Unbroken, Freedom, and Words Fail Me**.

No one would misread that *before* sentence. You could get away with it. But the second sentence floats straight into your mind. The words snugged up against the colon on the right (the book titles) follow directly from the word (*bestsellers*) on the left.

In the revisions below, notice again the difference that word position makes. Further edits would improve some of these sentences even more, but we'll stop here. We're focusing on the improvement that comes from bringing related words together.

> *Before:* There are three **choices** in this life: **be good, get good, or give up**. (Dr. House, *House, M.D.*)
>
> *After:* In this life, there are three **choices: be good, get good, or give up**.

> *Before:* These are the four most beautiful **words** in our common language: **I told you so**. (attributed to Gore Vidal)
>
> *After:* In our common language, these are the four most beautiful **words: I told you so**.

> *Before:* I like to knit with **yarns** that feel soft: **mohair, silk, and—softest of all—bamboo**.
>
> *After:* I like to knit with soft **yarns: mohair, silk, and—softest of all—bamboo**.

Meet Me at the Period (or Semicolon or Dash)

Where else could want-to-be-adjacents get together? At a period, where two sentences meet. The following example hooks up a pronoun (*it*) with its antecedent—the word that the *it* refers back to (*pizza box*)—at a period.

> *Before:* Frank picked up a discarded **pizza box**. The party had gone on for hours, and he was tired. He wanted to hit the sack. But **it** had made the apartment look messy. (The sack had made the apartment look messy?)
>
> *After:* The party had gone on for hours, and Frank was tired. He wanted to hit the sack. But he picked up a discarded **pizza box**. **It** had made the apartment look messy.

This trick also applies to independent clauses that *could* meet at a period but that you prefer to have meet, instead, at a semicolon or dash.

> Frank picked up the discarded **pizza box; it** had made the apartment look messy.
>
> Frank picked up the discarded **pizza box—it** had made the apartment look messy.

Meet Me at the Verb

To get a subject sidled up to its verb, nudge the intervening words out of the way. Hyphens (compound modifiers) and apostrophes (possessive case) can help.

> *Before:* The **plan** for doing the marketing via the website **is coming** together.
>
> *After:* The website-marketing **plan is coming** together.

> *Before:* The **foundation** of the old house **cracked**.
>
> *After:* The old house's **foundation cracked**.[115]

Meet Me at the End

Want your words to reach out and touch people? Get the right words to touch each other. When do you stop sliding words around? When you hit your deadline. Or when every last word has found its place.

---

115. Some "old line" authorities avoid adding 's to inanimate objects. They would keep *house* as an object of a preposition—*the foundation of the house*—rather than use the possessive—*the house's foundation*. This objection is waning. Most English speakers, like Bryan Garner, would find a house's ability to possess a foundation "generally unobjectionable" (Garner, *Garner's Modern American Usage*, 646).

Use Contrast:
The Long and Short of It

Writing has laws of perspective, of light and shade, just as painting does, or music. If you are born knowing them, fine. If not, learn them. Then rearrange the rules to suit yourself.
—Truman Capote in *Truman Capote: Conversations*

"What makes pages interesting? Why do people pay attention?" asks Jan White, an award-winning graphic designer and the author of a dozen books on visual techniques in publishing. "Because they sense something there that they are curious about—the subject-matter. But just presenting the subject in a take-it-or-leave-it way is not good enough," White says.[116] He uses a design element to motivate "lookers" and "turn them into readers." He uses it "to make the value of the message noticeable" and even make that message "so dramatic that it pops off the page irresistibly." Which design element?

Contrast.

The same technique that White uses to delight the eye, composer Franz von Suppé uses to delight the ear. His *Pique Dame Overture* begins with a series of whisper-soft, gently-paced notes that tiptoe from measure to measure until suddenly the orchestra explodes. The first time I heard this opening, I laughed—and wanted more.[117]

If the *Pique Dame Overture* were text, the opening measures would look like this:

---

116. Jan V. White, "Design 101: Contrast Makes the Difference," Xerox, accessed May 12, 2012, http://www.office.xerox.com/small-business/tips/contrast/enus.html.

117. To listen to this opening, visit "Pique Dame Ouvertüre," *Amazon*, accessed May 12, 2012, http://www.amazon.com/gp/product/B00138IMZI/ref=dm_mu_dp_trk5.

Deet. Deet. Deedle-deedle-deet deet. Deet. Deet. Deedle-dee-dle-deet deet. Deet. Deet. Deedle-deedle-deet deet. Deet. Deet-deet-deet. Deet-deet. Deet. Deet. Deedle-deedle-deet deet. Deet. Deet. Deedle-deedle-deet deet. Deet. Deet. Deedle-deedle-deet deet. Deet. Deet-deet-deet.

BLAST!

How can you emulate this kind of impact with words? Vary these two things:

- sentence length
- paragraph length

Look at the *Pique Dame* transcription, for example. Short, staccato sentences (*Deet*) alternate pleasingly with longer, more flowing sentences (*Deedle-deedle-deet deet*). Also, the long, gradually building first paragraph leaves us deliciously vulnerable to the second paragraph's compact blast.

Dramatically different sentence lengths, which make for the liveliest reading, elude writers who aren't thinking about sentence length. So, writer, think about it! Be bold. If you're going to contrast, *contrast*.

> **When you have a long sentence full of lots and lots and lots of words, like this one, put a shorter sentence—even a fragment—next to it. Like this.**

When you have a long sentence full of lots and lots and lots of words, like this one, put a shorter sentence—even a fragment—next to it. Like this.

The same goes for paragraph lengths: go for the drama. Of course, paragraphs still need to hang together based on their content. You still need transitions, topic sentences, examples, a logical progression of ideas—all the basics that you probably learned in high school. Few English teachers tell us, though, to use paragraphs as pigment. When you squint at your page, do you see a stack of same-sized gray rectangles as enticing as a cinder-block wall?

Yes?

No?

When it comes to deciding what goes in those eye-catchingly short paragraphs, follow Jan White's advice and "make the important elements stand out." Use the spotlight of white space. Only your truest fans read your long paragraphs. Everyone reads your short paragraphs.

Make them count.

Explore and Heighten:
Magic Words from a Playwright

Half the student's battle is learning basic skills, while the other half involves tapping into imagination, memory and a singular view of life and the world, a view no one else shares until you put it into words.

—ANNE BERNAYS, "PUPILS GLIMPSE AN IDEA, TEACHER GETS A GOLD STAR," *NEW YORK TIMES*

When a piece I'm writing needs a little more...something...I call to mind these five powerful syllables: explore and heighten. I owe this incantation to playwright Alan Gross, who had a group of us practically chanting it during a playwriting workshop I attended one summer during my college years. Whatever I'm writing, this phrase invariably nudges the content that oh-so-helpful bit further.

For example, while working at Nike World Headquarters, I discovered that my desk phone was made of material from recycled NFL helmets. Too perfect. *Got to tell friends about this*, I thought. I drafted a message:

Check it out—my Nike phone is made of old football helmets.

Then the magic words came to me as if the playwright himself were whispering them in my ear. *Explore and heighten.*

Phone...helmet...put one on, pull it down over my ears...am I wearing the phone?...this phone is my helmet...I am a formidable phone-calling foe...who is my opponent?...I hate being on hold...endless marketing hype...why can't they play some decent music?...waste of time...makes me want to hurt someone...

I returned to my message:

> Check it out—my Nike phone is made of old football helmets. Don't mess with me when I'm talking on this sucker. All those bleeping answering machines out there can put themselves on hold from now on. I make a call, it's going *through*.

One classic explore-and-heighten comes from a comedy sketch performed by two members of *Monty Python's Flying Circus*. John Cleese (playing Mr. Praline) walks into a pet shop, birdcage in hand, to complain to Michael Palin (playing the shopkeeper), that the parrot he bought a half-hour earlier is, and for some time has been, decidedly dead. Over and over Praline pleads his case. Over and over the shopkeeper insists, ridiculously, that the bird lives. For example, he claims that the bird is simply "pining for the fjords" of its native Norway. An exasperated Praline launches into a tirade of explored-and-heightened phrases:

> Look matey (*Picks up parrot.*) this parrot wouldn't go voom if I put four thousand volts through it. It's bleeding demised... It's not pining, it's passed on. This parrot is no more. It has ceased to be. It's expired and gone to meet its maker. This is a late parrot. It's a stiff. Bereft of life, it rests in peace. If you hadn't nailed it to the perch, it would be pushing up the daisies. It's rung down the curtain and joined the choir invisible. This is an ex-parrot.[118]

Well and good for Monty Python, you may think, *but the stuff I write involves a little less drama and far fewer dead birds.* I'm with you. But all of us spin out the occasional sentence or paragraph that's somehow lacking—thin or unclear or mundane. From time to time, we all

---

118. Joseph Black et al., eds., *The Broadview Anthology of British Literature: The Twentieth Century and Beyond*, vol. 6B, *From 1945 to the Twenty-First Century* (Toronto, ON: Broadview Press, 2008), 976. To see one of the many recorded performances of the Dead Parrot sketch, go to http://www.youtube.com/watch?v=4vuW6tQo218. For a summary of this sketch's history, entertaining in itself, see "Dead Parrot Sketch," *Wikipedia*, last modified May 27, 2012, http://en.wikipedia.org/wiki/Dead_Parrot_sketch.

write passages that threaten to make readers yawn or tilt their heads in puzzlement. Those are the times to add detail, the times to expand. Build up. Pile on the voom.

Invoke the magic of the mantra.

To give you an idea of how this mantra works for me, let me replay the Nike example in slow motion. I've drafted my original statement. I like its simplicity, but I don't want to send it out yet. I sense potential. I relax, breathe. Explore! My sentence, that so-so stretch of text, opens up. It becomes something palpable, something that I can crawl into. I feel it around my shoulders as I slip in, like a spelunker slipping into a small cave. (Caves and football don't usually mix, but bear with me. When we're thinking creatively, generatively, inventively, brainstorm-ingly, anything goes.) I'm in a wonderland of half-seen crannies and cavities and side chambers. This space is pure possibility. I look around, expecting—knowing—that discoveries lurk just out of view. Now, heighten! Phone ... helmet ... put one on, pull it down over my ears ... am I wearing the phone? ... this phone is my helmet ... I am a formidable phone-calling foe ... hate being on hold ... endless marketing hype ... why can't they play some decent music? (When Mick Jagger wrote "Satisfac-tion," did he see Muzak in its future?) ... waste of time ... makes me want to hurt someone ... who is my opponent? ... hate to be the one who gets between me and the person I'm calling ... snap on the chinstrap ... whose voice is that? ... sounds distinctly like my husband ... "Marcia! Get that phone on and get in there! Tell the quarterback to run dive-five-right. The left tackle pulls, blindsides the answering system, knocks it flat. Go straight through the line of clerks and into the secondary ... you'll pick up ten, twelve yards easy."

Explore and heighten. Of course this phrase possesses no magic in itself. It can't transform a thing. But it can usher you straight into your imagination, the place where *your* magic lives, where your ideas are born, where your power finds its source, where you will discover—waiting for you—the best sentences and paragraphs you will ever write.

Coming Soon
to the Small Screen

A well-crafted sentence matters more than ever in the digital age.
—*New York Times* online, *Draft* tagline

Are you reading this on a Droid? On an iPhone? On some other diminutive device being introduced even as I write this? Knowing that you might be gives me pause. The smartphone has become a primary reading device.[119] So unless you write nothing but lost-cat posters destined for telephone poles, or other print pieces that no one will ever upload to the web, you have little choice but to join me in grappling with this question: what must writers do differently to accommodate the small screen?

The answer, I believe, is … nothing.

Of course, mobile technologies require different formatting. We've all struggled to read a page intended for the big screen, pinching it and spreading it and pushing it around on a screen the size of a business card. When we then discover a mobile version of the site, or when an app like Wikipanion reformats it like magic, we love the ease of reading the same content optimized for that itty-bitty screen. A lot can be done with presentation (the way the information looks and acts).[120]

---

119. Industry veteran Maxwell Hoffmann tells technical writers, "The most common 'sheet of paper' for a first view of our 'documentation' will soon be that tiny screen held in your hand." See "#ICC12: Resizing Content for the Small Screen: Considerations," *Adobe Tech Comm Suite Evangelist* blog, February 24, 2012, http://blogs.adobe.com/mb hoffmann/2012/02/24/icc12-resizing-content-for-the-small-screen-considerations.

120. Marcia Riefer Johnston, "Marketing Pros: Time to Think Small," *Elliott Design* blog, April 11, 2011, http://elliottdesign.us/blog/marketing-pros-time-to-think-small.

But I'm not talking about presentation. I'm talking about the text itself. I'm talking about your words, which—willy nilly, sooner or later—will end up in someone's palm.

When it comes to text, today's writers (which is to say, writers for the small screen) must continue to do what writers have always done: cut, add, organize, and experiment. This chapter treats each in turn. But first, let's bust the myth that writers in the small-screen era should keep it short.

The Short-Sightedness of *Short*

Lots of people are talking about the need to write less—to *keep it short*—for mobile devices. Today's readers, we're told, suffer from "infobesity"; they "want less and less content."[121] We supposedly live in an "era of brevity," with our brains "rewired" in favor of short text.[122]

The problem is, *short* has no meaning. If I hand you a piece of string and say, "Make it short," where do you cut? You can't know. Who needs to do what with the string? Even when you know the answer, you don't want short. You want just right.

Think of writing as string cutting. Instead of aiming for short, aim for economical: the perfect length. Five words might be too many. Five thousand might be too few. The perfect length, even for the small screen, depends on "the context the reader brings," as technical communicator Tom Johnson points out.[123]

We need to ask the same questions today that writers have always needed to ask:

- Who will do what with the information?

- How much do they already know?

- What will they want to know (whether they know it or not)?

---

121. Robert Desprez, "Ruthlessly Edit When Writing for Mobile," *Robert Desprez Communications* blog, November 27, 2011, http://www.robertdesprez.com/2011/11/27/ruthlessly-edit-when-writing-for-mobile.

122. Tom Johnson, "Less Text, Please: Contemporary Reading Behaviors and Short Formats," *I'd Rather Be Writing* blog, January 21, 2011, http://idratherbewriting.com/2011/01/21/contemporary-reading-behaviors-favor-short-formats.

123. Ibid.

Until you do the work of answering these questions, you can't make good decisions about what to include and what to leave out, and any effort to streamline your content "*degenerates* into making sentences shorter"[124] (emphasis mine). So says JoAnn Hackos, who has spent decades teaching professionals about minimalist writing—writing that delivers no more *and no less* than what's needed.

Here's how Hackos sums up the difference between *short* and *minimal*:

> While reducing the volume and cutting the word count is certainly a desired outcome, it isn't the center of the minimalist agenda. Minimalist advocates understand that people do not want to read and actually do not read anything that does not appear to lead to fulfilling their immediate goals...The minimalist agenda focuses on usefulness and usability...[and requires companies to do] the harder and more time-consuming work of learning about customers.[125]

Writing that degenerates into a quest for shortness risks leaving readers puzzling over what Hackos calls "cryptic terminology" and "unnecessary brevity."[126]

Even Ernest Hemingway, famous for stories we call short, warns against seeking brevity for its own sake. His iceberg theory of writing makes clear that leaving things out requires discernment:

> If a writer...knows enough about what he is writing about he may omit things that he knows...The dignity of movement of an iceberg is due to only one-eighth of it being above water. A writer who omits things because he does not know them only makes hollow places in his writing.[127]

---

124. JoAnn T. Hackos, "What Makes Minimalism So Popular Today?" *CIDM Center for Information-Development Management News,* January 2008, http://www.infomanagementcenter.com/enewsletter/200801/feature.htm.

125. Ibid.

126. JoAnn T. Hackos, "An Application of the Principles of Minimalism to the Design of Human-Computer Interfaces," *Common Ground* (1999), 9:17–22. Reprinted by the Center for Information-Development Management, accessed March 1, 2012, http://www.comtech-serv.com/pdfs/Minimalism%20Human-Computer%20Interfaces.pdf.

127. Ernest Hemingway, *Death in the Afternoon* (New York: Scribner, 1932),

Legendary ad writer David Ogilvy goes so far as to claim that "long copy sells more than short." [128]

So when your inner guru (or your boss) chants, "Keep it short"—for mobile readers or anyone else—fire back (not necessarily out loud), "I'll make it as short as possible and as long as necessary."

Tips for Cutting

When you cut, do so more aggressively and empathetically than ever. As web-usability researcher Jakob Nielsen says, "Mobile use implies less patience for filler copy."[129] Do you tolerate filler copy in your writing? Tolerate no longer! Put on your drill-instructor hat, and pull up your text on a smartphone. Stare down every word. Do you contribute to the lineup, soldier? No? Then out you go!

Writer Maxwell Hoffmann attests to the value of reviewing his work on a smartphone. When he did this with a white paper that he had written several years before, he found parts of his text nearly unreadable. He reports, "My thumb nearly fell off scrolling through just three bulleted items ... I didn't have the patience for my own thoughts presented in the confines of a handheld smartphone."[130]

Instructive confession!

Want to recalibrate your definition of *concise*? Here's a suggestion. Make yourself a small-screen template in any word processor, or simply do some of your writing directly on a smartphone or on the back of a business card. That will get you thinking small.

This exercise may sound like unnecessary bother, but it helps you tighten your writing—for screens of any size. When I asked Hoffmann what percentage of his edits for the small screen also improve the reading experience on the big screen, he said, "One hundred

---

153–154; First Scribner e-book edition 2002, http://books.google.com/books?id= Wn69QsdwDlQC&printsec=frontcover&dq=death+in+the+afternoon.

128. David Ogilvy, *Ogilvy on Advertising* (New York: Vintage Books, 1985), 84.

129. Jakob Nielsen, "Mobile Content: If in Doubt, Leave It Out," *Nielsen Norman Group*, October 10, 2011, http://www.nngroup.com/articles/condense -mobile-content.

130. Hoffmann, "#ICC12: Resizing Content for the Small Screen: Considerations."

percent."[131] Readers *never* needed all those words. In this sense, author Luke Wroblewski's web-design mantra—"mobile first"—makes sense for writing too.[132]

> This 2-by-3-inch text box in my word processor approximates the size of a smartphone display. Typing into this box when I'm sitting at a big screen reminds me that readers might see only a handful of my words at a time. I recommend doing this exercise at least once, or else writing a few pages' worth directly on a smartphone. You'll return to the big screen thinking small. Oops— ~~running out of space!~~

A small-screen template.

Ultimately, though, as Ann Rockley (widely regarded as the "mother of content management") says, writers—along with illustrators and strategists and communicators of all disciplines—must consider "content first, *not* mobile first, eBook first, or any other 'first.'"[133] We write not for any one technology but for *any* technology.

The mandate to write tight transcends technology.

131. Maxwell Hoffmann, "Resizing Content for the Small Screen: Considerations for Single-Source Authoring for Tablet and Mobile Delivery" (lecture, #ICC12 Intelligent Content Conference, Palm Springs, CA, February 24, 2012).

132. By "mobile first," Wroblewski means, "Websites and applications should be designed and built for mobile first." Luke Wroblewski, *Mobile First* (New York: A Book Apart, 2011), 1.

133. Rockley says this repeatedly. See this interview, about fifteen minutes into the recording: "Designing Adaptive Content for a Mobile World," May 11, 2012, http://www.contentrules.com/videos/#annrockley. See also Rockley and Cooper, *Managing Enterprise Content*, 136–139.

Tips for Adding

Let's say that you now regularly cut every bit of text you can. Don't stop there. Remember the complementary necessity: adding. Small screen, schmall screen. Go after thin copy. As author William Zinsser advised back when typewriters ruled, "strip your writing down" and then "build it back up."[134] Ask yourself, *What else would readers appreciate?* Your readers' fingers are flicking for a reason. They're not chasing after copy that's skimpy. They want copy that answers their questions, copy that's coherent, logical, useful, convincing, relevant, inspiring, edifying, amusing, fresh, clarifying, and sufficiently detailed.

What kind of details could you add? Choose from paragraph-development standbys like these:

- examples

- anecdotes

- definitions

- quotations

- metaphors

- instructions

- descriptions (concrete language: sights, sounds, tastes, smells, sensations)

- answers to the *W-H* questions (who, what, when, where, why, whether, how, how much, how many, how often, what else you got)

Some businesses find their fortunes by providing the right details. Kyle Wiens, cofounder of iFixit—a popular website that provides free repair manuals and advice forums—sums up the payoff this way: "Users will love people who teach them what they want to know."[135] The iFixit website serves up detailed procedures that display beautifully on a screen of any size. The iFixit folks and their engaged customers add,

---

134. Zinsser, *On Writing Well*, 20.

135. Kyle Wiens, "From Web to iPhone to Android to iPad: The iFixit.com Story" (lecture, #ICC12 Intelligent Content Conference, Palm Springs, CA, February 24, 2012).

add, add relevant information, every day, to this large website. As a result, they draw large, loyal audiences who order a lot of parts and tools from them.

How has this worked out? Next to Apple, iFixit sells more Apple parts than anyone in the world.

Era of brevity? The success of information-rich websites like iFixit gives the lie to such labels.

A repair procedure as viewed in the iFixit app on a smartphone.

Cut and *add* go together like *diet* and *exercise*. Do both, as disciplined writers always have, and watch each flabby sentence and sagging paragraph transform into a specimen. A hunk. A perfect 10. Take the iFixit example. Deliver writing like that, and your readers—the ones who are out there right now looking for exactly what you have to offer—won't be able to pull their eyes away, no matter what size the screen.

Tips for Organizing
We writers for the wee screen need to organize our material more carefully than ever, especially when it comes to what Nielsen calls

"complicated content," like "nightmarishly long" documents.[136] Nielsen says that complicated content is "roughly twice as hard to understand" on a smartphone ("a peephole") as on a desktop monitor. Nielsen claims that "a smaller screen hurts comprehension" because people can see less at one time and because they have to move the page around more. He recommends "adding structure and navigation" to "create a tight information space."

Good advice, if not new. Writers have been pursuing these goals— structure, navigation, and tight information spaces (to stretch the terms)—since the days of hieroglyphics. These goals have less to do with technology than with the down-in-the-bones organization of the material. Aside from formatting, any problems with complicated and nightmarishly long content lie in the content itself. The small screen may double the nightmarishness, but it doesn't create it. Nightmarish in, nightmarish out.

So let's unconflate technology issues and writing issues. Let's unhook technology from this part of the discussion and review a few timeless practices for organizing information.

- **Select the right info.** Determine who needs what information; resist any temptation or pressure to burden your material with extras.

- **Chunk it.** Divide the information into logical sections. For more on chunking information—for web pages or anything else—see the seminal book on information architecture by Louis Rosenfeld and Peter Morville: *Information Architecture for the World Wide Web.*[137]

- **Label it.** Assign each section a tight, clear subheading to help people find their way through the information.

---

136. Jakob Nielsen, "Mobile Content Is Twice as Difficult," *Nielsen Norman Group*, February 28, 2011, http://www.nngroup.com/articles/mobile-content-is-twice-as-difficult.

137. Louis Rosenfeld and Peter Morville, *Information Architecture for the World Wide Web* (Sebastopol, CA: O'Reilly, 1998), especially "Organizing Information" (22–46) and "Grouping Content" (147).

- **Arrange it.** Place the sections in a useful order, adding cross-references or links to indicate the most important nonlinear relationships. See my blog post, "Organizing Hard So Information Is Why?" (http://howtowriteeverything.com /organizing-hard-so-information-is-why).

Wikipedia serves as an ideal source of examples. Many Wikipedia pages exhibit informed selecting, logical chunking, clear labeling, and meaningful arranging. These old-as-the-pyramids organizing practices result in pages that adapt beautifully to screens big and small.

Organize your material with care, as this Wikipedia writer did, and no one will ever call your writing nightmarishly long, no matter how long the piece or how small the screen. Organized in, organized out.

On this Wikipedia page,[138] the same subheadings that act as visual skimming aids on a big screen (left)—Etymology, Types, etc.—act, on a smartphone screen (right), as an interactive advance organizer (list of contents). Whether readers navigate this page with their fingertips or with their eyes, they owe their positive experience (if they happen to care about such things as deponent verbs), in large part, to the writer's organizational skills.

Tips for Experimenting

All kinds of issues come up as you look at your text on a tiny screen. You have plenty of opportunity to experiment. If your documents, for

---

138. "Participle," *Wikipedia*, last modified June 24, 2012, http://en.wikipedia.org /wiki/participle.

example, require annotated illustrations (text callouts) or complex tables, consider converting them to a fixed-layout format, which can work surprisingly well on a small screen. Fixed-layout formats are evolving so fast that I hesitate to say much about them except that they fill a need. They enable the publication of cookbooks, comic books, children's books, and technical manuals, in which images and text (or text and text) must not reflow but stay connected to make sense.

For example, Bruce Ashton, a fellow attendee of the 2012 Intelligent Content Conference, showed me a fixed-layout handbook on his iPhone. He explained his company's luck: the toolbox-size handbooks they'd been printing for thirty years had just the right proportions for smartphones. Converting the handbooks to a fixed-layout format "is not a quick process but it works," Ashton writes.[139] This format enables his company, IPT Publishing & Training Ltd., to deliver complex tables and illustrations legibly on an iPhone. (As of this writing, the fixed-layout format is incompatible with other smartphones.)

As you encounter challenges in writing for screens of all sizes, do what innovative writers have done since obsidian first scratched sandstone: let the spirit of experimentation lead you.

Summary: The More Things Change…

When you write for the small screen—which is to say, when you write— you can't go wrong if you rededicate yourself to the basics: cut, add, organize, and experiment. Ignore the hollow advice to keep it short. Lose the stereotype of mobile users as rushed and desktop users as tolerant of "happy talk."[140] Tune out those who advise you to create separate mobile "lite" versions of your content. Listen to user-experience pro and content strategist Karen McGrane, who says, the "recommendation that mobile sites should cut content and features relegates [mobile-only] users to second-class citizens."[141] Listen to interaction

---

139. Bruce Ashton, e-mail message to the author, March 1, 2012.

140. Stephanie Rieger, "Mobile Users Don't Do That," *Beyond the Mobile Web* blog, February 10, 2012, http://stephanierieger.com/mobile-users-dont-do-that.

141. Tanya Combrinck, "Designers Respond to Nielsen on Mobile," *.net* blog, April 12, 2012, http://www.netmagazine.com/news/designers-respond-nielsen-mobile-121892.

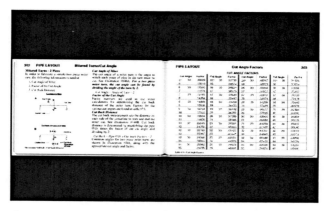

A zoomed-out two-page spread from a fixed-layout e-book. This example comes from a free excerpt of "IPT's Pipe Trades Handbook" by Robert A. Lee, downloaded from the iBookstore. Pages reproduced with permission. This screen shot was taken on an iPhone.

PIPE LAYOUT   **Cut Angle Factors**   303

CUT ANGLE FACTORS

Cut Angle	Factor	Cut Angle	Factor	Cut Angle	Factor	Cut Angle	Factor
5° 30	.09629	15° 30	.22732	25° 30	.46797	35° 30	.71329
6	.10510	16	.28674	26	.48773	36	.72654
6 30	.11393	16 30	.29621	26 30	.49858	36 30	.73996
7	.12278	17	.30573	27	.50952	37	.75355
7 30	.13165	17 30	.31530	27 30	.52057	37 30	.76733
8	.14054	18	.32492	28	.53171	38	.78128
8 30	.14945	18 30	.33459	28 30	.54295	38 30	.79543
9	.15838	19	.34433	29	.55431	39	.80978
9 30	.16734	19 30	.35412	29 30	.56577	39 30	.82424
10	.17633	20	.36397	30	.57735	40	.83910
10 30	.18534	20 30	.37388	30 30	.58904	40 30	.85408
11	.19438	21	.38386	31	.60086	41	.86929
11 30	.20345	21 30	.39391	31 30	.61280	41 30	.88472
12	.21256	22	.40403	32	.62487	42	.90040
12 30	.22169	22 30	.41421	32 30	.63707	42 30	.91633
13	.23087	23	.42447	33	.64941	43	.93251
13 30	.24008	23 30	.43481	33 30	.66188	43 30	.94896
14	.24933	24	.44523	34	.67451	44	.96589
14 30	.25862	24 30	.45573	34 30	.68728	44 30	.98285
15	.26795	25	.46631	35	.70021	45	1.0000

Table #75 - Cut Angle Factors

A zoomed-in look at the table text, as legible as you please on an iPhone. The usefulness of these well-organized pages calls into question Jakob Nielsen's claim that the small screen hurts comprehension.

designer Josh Clark, who says, "Stripping out content from a mobile website is like … stripping out chapters from a paperback just because it's smaller. We use our phones for everything now; there's no such thing as 'this is mobile content, and this is not.'"[142]

---

142. Josh Clark, "Nielsen Is Wrong on Mobile," *.net* blog, April 12, 2012, http://www.netmagazine.com/opinions/nielsen-wrong-mobile.

You don't need a double standard to write for small screens. Good writing is good writing. Think small, yes. But don't stop short just because the screen does. r y mght s wll jst dlt ll th vwls.[143]

---

143. Or you might as well just delete all the vowels.

How Not To Do How-To

Cheshire Cat: You know? We could make her really angry!
Shall we try?
Alice: Oh, no, no!
Cheshire Cat: Oh, but it's loads of fun!
—DISNEY'S *ALICE IN WONDERLAND*

My Amazon Kindle Fire e-book reader blows my mind. It holds a fathomless supply of books, magazines, and movies—and it takes up less room in my purse than my makeup kit. The interface thrills like a party trick; if I had a library's worth of e-books sitting on this infinite bookshelf, the gentlest finger stroke would send their bright covers flying off the edge of the screen like cards in a game of 52 bazillion pickup.

The user guide, I'm afraid, blows my mind too. It's among the least helpful product manuals I've ever encountered.

The step-by-step information in this user guide—the information that most new-product users want—has two baffling problems: the steps are camouflaged in dense paragraphs, and, more dismaying, the steps describing certain e-book features don't work in this e-book!

How can the preeminent e-book-reading device come with such a stunningly *not* preeminent introduction to e-book reading?

I turn on my new toy, and there's the cover of the user guide. It's a pleasing, tightly packed arrangement of the letters K, I, N, D, L, E in various sizes and styles. These wooden blocks of movable type, facing every which way in attractively juxtaposed shades of rust and pewter, say, "Farewell, old ways." I'm enchanted. I can't wait to embrace the new.

I touch the cover, and the book opens. I flick through a few pages, looking for the how-to part. I come to a picture of the main Kindle screen. Yes, I expect that image, those callouts. I keep flicking. Paragraphs, more paragraphs. Where are the procedures? Where are the instructions? I want to know what I can do and how to do it. I comb the text for the 1-2-3s.

In the 60-some flicks that it takes me to reach the end of the book, I come across one set of numbered steps. All the other instructions—which make up at least half of the book—were left inexplicably unnumbered. Even more exasperating, they are strung together into soporific paragraphs (zzzz ...) like this one:

> To add a note, press and hold on a word, or press and drag to select multiple words. When the contextual menu appears, tap Note and use the onscreen keyboard to type your note. To highlight a passage, press and hold until the magnifying box appears, then drag your finger to the last word you want to highlight and release. When the contextual menu appears, tap Highlight. Your Kindle Fire saves your place in whatever content you're reading, or you can manually add a bookmark. To add a bookmark, tap the screen, then tap the bookmark icon at the top right of the screen. To delete a bookmark, tap it. To view your notes, highlights, and bookmarks, tap the screen to bring up the Options bar, then tap the Menu icon.

Really, Amazon? One unillustrated blob of text for all these tasks? Why force readers to hack their way through such a thicket of instruction? Did you, with your reputation for constantly testing your website for usability, do no usability testing on this document?

Determined to learn what I can, I pick an action: "Tap the screen, then tap the bookmark icon." I tap the screen. No bookmark icon appears. I tap again. And again. I feel stupid. What am I doing wrong? After a few confusing failures, an unlikely possibility occurs to me: this e-book *does not support the e-bookmark function it describes.*

No way.

(*Note to self:* Remember this defeated, I'm-stupid feeling. Nothing motivates writers to clarify and refine like empathy for frustrated readers.)

I tap again, holding my finger down longer this time, making a deal with the page: I give you extra touch time, you give me that bookmark icon. Eventually, denial and bargaining give way to reluctant acceptance.

Way.

I move on to the descriptions of note-taking and highlighting, only to discover that neither of these functions works here either. The instructions might be accurate, but I can't use them in this e-book of a user guide. Amazon, where were your user-experience designers and testers when it came to users' experience of the guide that bears their name?

(Remember this feeling!)

I slog on. I tease the tasks apart myself—can't help it. Headings and numbered steps emerge. I envision a makeover:

> I tap again, holding my finger down longer this time, making a deal with the page: I give me extra touch time, you give me that bookmark icon. Eventually, denial and bargaining give way to reluctant acceptance.

To Add a Note

1. Press and hold on a word.

2. (Optional) Drag either handle to highlight a group of words.

3. Tap Note.

4. Type your note.

Pssst. Don't try these steps here; they don't work in this book.

To Highlight a Passage

1. Press and hold on a word.

2. (Optional) Drag either handle to highlight a group of words.

3. Tap Highlight.

Pssst. Don't try these steps here ...

To Add a Bookmark

1. Tap the screen.

2. Tap the bookmark icon at the top right of the screen.

Pssst. Don't try these steps here ...

Okay, I've vented. I feel better now. In fact, I feel good. Exhilarated even. Admittedly, I'm a user-guide junkie. When I buy anything, I read the instructions. If a user guide is printed, I scribble notes in its margins. I like to understand how things work. I like to understand how *information* about things works. I get a kick out of thinking about how to do how-to. It tickles my brain to analyze explanations: those that work and—especially rewarding—those that don't.

I *heart* makeovers.

Come to think of it, Amazon ... thanks for the good time! Let's do this again. Call me?

How To Do How-To:
Watch Your Steps

A-B-C. It's easy as 1-2-3. As simple as do-re-mi, A-B-C, 1-2-3, baby you and me girl.
—THE JACKSON FIVE, "ABC"

If you've ever jotted down directions to get to someone's house, make a soufflé, use an e-book reader, repair an engine, or transplant a heart, you've written a how-to, also known as a procedure. Maybe you consider yourself a technical writer, in which case you may think you already know what you need to know about writing procedures. Maybe you consider yourself a nontechnical writer, in which case you may think you don't need to know *anything* about writing procedures.

Whatever kind of writer you consider yourself, I hope that you give this topic a chance. You never know when you'll be called on to explain how to do something. When you write a helpful procedure that makes its way to the right person in the right way at the right time, you make that person's life better.

> When you write a helpful procedure that makes its way to the right person in the right way at the right time, you make that person's life better.

A note on terminology: Technical writers often differentiate, with good reason, between *procedures*—instructions that walk people through product functions—and *task-oriented instructions*—instructions that help people accomplish their goals, which might not directly correspond to product functions. For my purposes here, I use the term *procedure* broadly to describe step-by-step instructions of any kind.

A note on sources: The step-writing tips in this chapter come from a variety of sources, including writers I've known and books that you'd never hunt down if I gave you their ISBN numbers. You couldn't pry the titles of these books out of me even with rapturous declarations of fondness for my footnotes.[144]

Overview

You may wonder why this chapter goes on for so long since procedure writing consists of, basically, *do this, do that*. Not much to it. But when you write steps, just as when you take steps, you can fall into traps. To help you avoid those traps, this chapter describes the following strategies:

- Get clear on what a step is.

- Eliminate steps that lack action.

- Number only steps that must occur in sequence.

- Put numbered steps in the right sequence.

- Create subtasks (with substeps).

- Flag any optional, conditional, or branching steps.

- Rewrite ambiguous steps.

- Title the procedure.

After you've mastered these strategies, bring on that heart-transplant manual.

Note: For long pieces of writing, like this chapter, readers deserve a tell-'em-what-you're-gonna-tell-'em section (also known as an overview, an executive summary, an abstract, a preview, a *précis*, or an advance

---

144. If you insist, I'll divulge my favorite sources on procedure writing:
- Sun Microsystems, Inc., *Read Me First! A Style Guide for the Computer Industry*, 2nd ed. (Upper Saddle River, NJ: Prentice Hall, 2003), 127–132.
- Gretchen Hargis et al., *Developing Quality Technical Information: A Handbook for Writers and Editors*, 2nd ed., IBM Press (Upper Saddle River, NJ: Prentice Hall, 2004), 41–44.
- Kurt Ament, *Single Sourcing: Building Modular Documentation* (Norwich, NY: William Andrew Publishing, 2003), 123.
- JoAnn T. Hackos, PhD, and Dawn M. Stevens, *Standards for Online Communication* (New York: Wiley, 1997), 271–272.

organizer). Research shows that advance organizers have "measurable benefits."[145] Textbook author Mike Markel says that advance organizers "improve coherence by giving readers an overview of the discussion before they encounter the details."[146] I recommend including advance organizers, whatever you call them (with or without bullets), in your own long pieces. To aid recognition, use the same terms in the advance organizer as you use in the piece itself. For example, the first bullet point above uses the same wording as the heading you're about to read.

Get Clear On What a Step Is

A step is an action. A numbered step is an action in sequence. So far, so good.

Eliminate Steps That Lack Action

A step that includes no action is not a step. In the example below, Steps 1–4 are steps. Step 5 is a point of information, not a step. It doesn't tell the cook to do anything. Every step must include a verb in the imperative mood (also known as a bossy verb), like *boil*, *make*, or *bake*. No imperative verb, no step. To fix Step 5, you'd delete its number and put the information where it belongs.

Before:

1. Boil and drain the cabbage.

2. Make the white sauce.

3. Layer the cabbage, the white sauce, and the cheese.

4. Bake at 350°F until the casserole bubbles.

5. **This dish won't take long to bake—maybe 15 minutes— if the cabbage and sauce are hot when you layer them.**

---

145. William Lidwell, Kritina Holden, and Jill Butler, *Universal Principles of Design: 125 Ways to Enhance Usability, Influence Perception, Increase Appeal, Make Better Design Decisions, and Teach through Design—Revised and Updated* (Beverly, MA: Rockport, 2010), 18.

146. Markel, *Technical Communication*, 253.

After:

1. Boil and drain the cabbage.

2. Make the white sauce.

3. Layer the **hot** cabbage, the **hot** white sauce, and the cheese.

4. Bake at 350°F **for 15 minutes** or until the casserole bubbles.

Another kind of step that lacks action, like Step 5 below, is a result. Results—if they need to be stated at all—belong right in the step.

Before:

4. Bake at 350°F for 15 minutes.

5. The top becomes golden brown.

After:

4. Bake at 350°F for 15 minutes. **The top becomes golden brown.**

Number Only Steps That Must Occur in Sequence

Steps that can be done in any order (or single steps) need no numbers. Use bullets, or simply delete the numbers.

Before:

1. Pour yourself a glass of wine.

2. Take the cabbage out of the fridge.

3. Put on a chef's hat.

After:

• Pour yourself a glass of wine.

• Take the cabbage out of the fridge.

• Put on a chef's hat.

Put Numbered Steps in the Right Sequence

Writers sometimes stick a numbered step at the end because they thought of it at the end. (Doubt me? While writing this chapter, I received an e-mail describing, step-by-step, how to apply the latest update of a certain smartphone app. At the bottom of the procedure came this note: "We advise you to sync your library before applying the latest app update." Too late!)

Before:

1. Boil and drain the cabbage.

2. Make the white sauce.

3. Layer the hot cabbage, the hot white sauce, and the cheese.

4. Set the oven to 350°F, and wait for it to get hot.

5. Bake for 15 minutes.

Note: To save time, turn on the oven earlier.

After:

1. Preheat the oven to 350°F.

2. Boil and drain the cabbage.

3. Make the white sauce.

4. Layer the hot cabbage, the hot white sauce, and the cheese.

5. Bake for 15 minutes.

Create Subtasks (With Substeps)

When multiple steps work together to accomplish something, clue the reader in by creating a subtask. For example, in the *before* example below, Steps 2, 3, and 4 tell how to make a white sauce, but they don't spell that out. The *after* example does. Readers understand, at a glance, that they have not five but three main things to do.

Before:

1. Boil and drain the cabbage.

2. **Melt the butter in a small saucepan.**

3. **Add the flour and whisk the mixture until it becomes frothy.**

4. **Gradually add the milk, stirring constantly over low heat.**

5. Layer the hot cabbage, the hot white sauce, and the cheese.

After:

1. Boil and drain the cabbage.

2. **Make the white sauce.**

    a. Melt the butter in a small saucepan.

    b. Add the flour and whisk the mixture until it becomes frothy.

    c. Gradually add the milk, stirring constantly over low heat.

3. Layer the hot cabbage, the hot white sauce, and the cheese.

Flag Any Optional Steps

An optional step can be ignored at the reader's discretion. Flag any optional steps by leading with the word *optional*, as in Step 5 below.

Example:

4. Bake at 350°F for 15 minutes.

5. **(Optional)** Broil for 2 minutes.

6. Remove the casserole.

Avoid starting optional steps with *if* (as in "If you want to ..." or "If desired ..."). *If* lead-ins are more suitable for conditional steps.

Flag Any Conditional Steps

A conditional step can be ignored only under certain circumstances. Flag any conditional steps by leading with the word *if*, as in Step 2 below.

Example:

1. Boil and drain the cabbage.

2. **If no one has wheat allergies**, make the white sauce.

3. Layer the hot cabbage, the cheese, and, if you're using it, the hot white sauce.

In a conditional step, you don't need to say, "Otherwise, skip to the next step." If you need an *otherwise*, you're probably looking at a branching step.

Flag Any Branching Steps

A branching step describes alternative ways to accomplish an action. The person following the steps selects the appropriate branch (alternative). Typically, writers flag the branches with a bulleted list or a table.

Branches in a bulleted list:

5. Set the oven to medium heat.

   - **If your oven uses a Fahrenheit scale, set it to 350°.**
   - **If your oven uses a Celsius scale, set it to 175°.**
   - **If your oven uses a Kelvin scale, set it to 450°.**

Branches in a table:

5. Set the oven to medium heat.

Scale	Medium
Fahrenheit	350°
Celsius	175°
Kelvin	450°

If the same branches recur throughout a procedure, make separate procedures. Say you're writing a pizza recipe that says—over and over—in your own kitchen do this, at a campfire do that. You might as well create two standalone recipes. (The campfire recipe has one step: order pizza.)

Tip: Avoid false branching steps, which include branches like "If yes, skip to the next step." This kind of step, as shown in Step 6 below, is an optional step in disguise.

> *Before:*
>
> 6. Determine whether your counter resists heat.
>   - **If yes, skip to the next step.**
>   - **If no, place hot pads on the counter.**
> 7. Remove the casserole from the oven, and place it on the counter.

> *After:*
>
> **6. (Optional) Place hot pads on the counter.**
> 7. Remove the casserole from the oven, and place it on the counter.

Rewrite Ambiguous Steps

Ambiguous steps leave you wondering, *Am I supposed to do this step, or can I skip it?* Ambiguous steps often start with *to*. Delete the *to*. If the step is optional, say so.

> *Before:*
>
> 8. **To flambé** the casserole, douse it with liquor, and blast it with a blowtorch.

> *After:*
>
> 8. **Flambé** the casserole by dousing it with liquor and blasting it with a blowtorch.
>
> *or*
>
> 8. **(Optional) Flambé** the casserole by dousing it with liquor and blasting it with a blowtorch.

Title the Procedure

Give each procedure a title.

> **Tip 1: Choose a capitalization style, and stick with it.**
>
> Create a Casserole Flambé (title case)
>
> Create a casserole flambé (sentence case)

Warning: Title-capitalization questions—*with? With?*—can lead to bloodshed. If you work with a team of writers, let a style guide serve as referee. First, of course, you must agree on a style guide.

> **Tip 2: Choose a syntax, and stick with it.**
>
> How Do I Bake a Casserole Flambé? (question form)
>
> How To Bake a Casserole Flambé (how-to form)
>
> To Bake a Casserole Flambé (infinitive form)
>
> Baking a Casserole Flambé (gerund form)
>
> Bake a Casserole Flambé (root-verb form)

The question form works well if you have plenty of space for title text and if the reader may feel anxious about the topic. A pamphlet describing a surgical procedure, for example, might walk the prospective patient through a set of questions as if the writer were reading the concerned patient's mind.

The how-to lead-in may lend itself to the title of a magazine article or…I don't know…a book, say, or the odd book chapter. *How to* draws readers. (Here you are.) If you're writing a handbook or website full of procedures, though, and if their titles will appear together in a topic list or in search results, all those stacked up *How to*'s amount to pure clutter. In that case, drop *How to*. People browsing your titles on smartphones, especially, will thank you.

The infinitive (*to*) form shortens a *how to* title by four characters. Some tech writers use the infinitive form to create a stem sentence within a procedure. A stem sentence is a subheading that flags the break between any introductory paragraphs and the first step.[147] In procedures that include stem sentences, the overall procedure title needs a distinct form, often a gerund.

The gerund (an *-ing* word that acts as a noun[148]) is waning in popularity for procedure titles, and for good reasons. Compared with the root verb—we're getting to root verbs—the gerund takes up more space, takes longer to process cognitively, and raises more questions for translators, who often translate it the same way as a root verb anyhow.

When in doubt about what kind of title to choose, use the verb in root form, an imperative look-alike but not a command. Like the gerund, this form enables you to put the action word first. The compact root form has the added advantage of revealing more of the title on a small screen. Finally, because people usually type root verbs in search boxes, this form of title shows up more reliably in search results.

Warning: Title-syntax questions, like capitalization questions, can lead to bloodshed. Say what you will about style guides, they help keep the peace.

Take the Last Steps

When you've finished writing your procedure, put it in front of people.

1. Observe.
2. Edit.
3. Repeat.

---

147. For a discussion of stem sentences and of some reasons that writers might want to let go of them, see Carol Geyer, "Stem Sentences," DITA.XML.org: Online Community for the Darwin Information Typing Architecture OASIS Standard, March 20, 2008, http://dita.xml.org/wiki/stem-sentences.

148. Not all *-ing* nouns are gerunds. See the glossary under "*-ing* noun" on page 202.

Your Words Come Alive with a Hint of Music

*To me, the greatest pleasure of writing is not what it's about,
but the inner music the words make.*
— TRUMAN CAPOTE IN *McCALL'S*

Let's leave the essay form behind
For just a beat or two.
(Go easy, please. The Bard I'm not!
Here, only a poem will do.)

Form must follow function,
And this form, although demanding,
Gives us access to a different
Kind of understanding.

Things that are worth saying need
To say more than they say.
By this I mean that sound and rhythm
Have a role to play.

You might not write poetry.
You might not use rhyme.
But elements of poetry
Make common words sublime.

Struggling with a title
As appealing as cold trout?

A bit of alliteration
Makes it jump right out.

Double up a consonant:
M-M, B-B, L-L.
"Tea for the Tillerman"—
Cat Stevens does it well.

Invite your readers in
With a title they can hum:
"Re-Thinking In-Line Linking"[149]—
Look out, Nelly! Here they come!

Struggling with a clincher?
Want to strike a chord?
Inject a subtle cadence,
And—voilà—you are adored.

Take a tip from Shakespeare,
Who sent Hamlet offstage rhyming.
People pay attention
When your tagline swings with timing.

Jotting down a list of things
Or series of like phrases?
Keep the structure parallel:
All nouns, all verbs, all whatever as long as no clunkers break
    the pattern.

If music be the food of love,
Let it feed your writing too.

---

149. Mark Baker, "Re-Thinking In-Line Linking: DITA Devotees Take Note!" *The Content Wrangler* blog, May 3, 2012, http://thecontentwrangler.com/2012/05/03/re -thinking-in-line-linking-dita-devotees-take-note.

Make the hills come alive with the vibe of your syllables
Ricocheting through.

Channel Theodor Geisel
(The playful Dr. Seuss).
Pentameterize an iamb or two,
Turn your writing loose!

Tap your toes, and you will find
That your readers want to stay.
Shape and tune till it sounds just right—
You'll take their breath away.

It doesn't matter what you write.
You can keep your line breaks random.
But you'll never regret daring to set
Meaning and "music" in tandem.

(Don't write like this—good heavens.
This is pure exaggeration.
Give sentences and paragraphs
Just a hint of syncopation.)

You get the point, no need to make
This poem any longer.
I'll sum it up. Here's how to make
Strong writing even stronger:

Jazz your phrasing up a lick—
Dull writing is a bummer.
Fine tune your inmost inner ear.
Release your inner drummer.

PART III

Up with Writing

I'm just going to write because I cannot help it.
—ATTRIBUTED TO CHARLOTTE BRONTË

The Importance of Re-Vising

The difference between writing that works and writing that doesn't work is very small adjustments. I've discovered the way to figure out what those adjustments are is to spend time on [the writing] ... to look at it again and again.
—ANDREW CLEMENTS IN THE *WASHINGTON POST*

"It's hard to keep all the rules in mind while I'm writing," says my daughter, Elizabeth, who blogs about her Peace Corps experiences.

Keeping all the rules in mind is not hard. It's impossible.

Well then, what should be happening in your mind as you write? My answer involves seeing and seeing and seeing again. But before this answer can make sense, I have to address a bigger question: how does the creative mind work?

This line of inquiry may seem more philosophical than practical. But don't let it scare you off. This exploration of creativity leads me to practical insights. I sum them up at the end. *Take these insights to heart, and they'll do more for your writing than the rest of the chapters combined.* Skip to the end if you like, but if you do, you won't fully appreciate the word *build* in this book's subtitle. You also won't know what I mean by *re-vising*. To explain what I mean by that hyphenated term, which you won't find in any dictionary, first I review some research into the design process of architects, and then I connect that research to my experience with creativity in writing—all writing, that is, not just so-called creative writing.

Essentially, this chapter gives you an extended definition of *re-vising*, and then it tells you why you might care and what you can do about it.

How Does an Architect's Creative Mind Work?

To explore the way the creative mind works, you have only to observe architects as they sketch designs. Gabriela Goldschmidt, who wrote the 1991 article "The Dialectics of Sketching," did just that.[150] Goldschmidt analyzed the comments that architects made as they sketched during the early stages of design. She discovered that the sketchers' thoughts followed a "pendulum pattern," a "swaying movement" between two ways of seeing: "seeing as" and "seeing that."

She developed a way of depicting this cycle, this "ping-pong pattern" between *as* and *that* modes of seeing. Her depictions look something like this:

Seeing as, seeing that, seeing as, seeing that.

Goldschmidt records in detail the experiences of one particularly creative architect. When he sees *as*, something in the sketch reminds him of another *thing*. For example, as he sketches a library, he sees a group of shapes *as* a puzzle, a series of aligned items *as* an axis, a certain spot *as* a dome.

On the other hand, when the architect sees *that*, he uses what Goldschmidt calls a "nonfigural" eye. He sees in terms of "generic architectural rules" or "design values and beliefs." He may see *that* a building needs a strong relationship to its site, or *that* trees would provide shade, or *that* an atrium would fit in.

Each way of seeing—seeing *as* and seeing *that*—informs the other. Back and forth. Goldschmidt gives this dialectic various names: "design reasoning," "visual thinking," "sketch-thinking."

In the world of sketching, the process looks like this: The architect erases. He adds squiggles. He pauses. He erases again. Insight by

150. Gabriela Goldschmidt, "The Dialectics of Sketching," *Creativity Research Journal* 4, no. 2 (1991): 123–143. Read the abstract, preview the first page, or purchase the whole article here: http://dx.doi.org/10.1080/10400419109534381.

insight—seeing as, seeing that, applying what he knows about how to make buildings work for people, applying what he knows about *anything*—he ushers his rendering of library and grounds to a point of satisfaction.

What Does Sketch-Thinking Have to Do with Writing?

I believe that sketch-thinking—the dialectic that drives the masterful designing of a building— also drives the process of writing. Goldschmidt agrees: "I can easily see how 'playing with words' can yield the equivalent of seeing as and that."[151]

> **When I see *as*, I see in metaphors. When I see *that*, I see in grammar rules and style guidelines.**

While I'm writing, when I see *as*, I see in metaphors. When I see *that*, I see in grammar rules and style guidelines.

Example:

I write: *This statement is an example of itself.*

I see **that** the *is* violates my guideline of avoiding *is*.

I revise: *This statement exemplifies itself.*

I see the statement **as** a hand drawing itself.

I revise: *This statement, Escher-like, embodies itself.*

I see **that** the statement bores me.

I delete the whole thing.

Two ways of seeing. To write powerfully, you need them both.

When I'm most absorbed in writing—when I'm in that focused, energized, productive, intrinsically rewarding state of mind known as flow—I oscillate constantly between seeing *as* and seeing *that*. I don't flip at will from one kind of seeing to another and then back again, as in "Time to put on my seeing *that* glasses now." But I do look at the same

---

151. Gabriela Goldschmidt, e-mail message to the author, February 24, 2011.

words, sentences, and paragraphs over and over, sometimes seeing *as*, sometimes seeing *that*.

This kind of oscillation exacts a price. When I revise and revise—or, as I think of it, re-vise and re-vise (as in re-see; the Latin *revisere* means "to look at again")—I pay for the crafting in minutes or hours or days. Not that I regret investing that time; what else could gratify me so much in the doing and leave me with something of such value when I'm done? As William Zinsser says, referring to a sentence that took him almost an hour to write, "No writing decision is too small to be worth a large expenditure of time. Both you and the reader know it when your finicky labor is rewarded by a sentence coming out right."[152]

Advice That Can Do More for You Than the Rest of the Chapters Combined

As promised, I'm wrapping up with some advice.

- You can't do much re-vising when you're crunching on a deadline or plowing through hundreds of e-mails. But when you're writing something important, or when you seek the fullest satisfaction that working with words can bring, set aside time—lots of time.

- Just as architects must know a lot about buildings (and people and life), writers must know a lot about language (and people and life). Every bit of knowledge has the potential to help you re-vise your words. Learn everything you can about everything you can.

There you have it. Take your time, and fill your head. As you do, you'll see new possibilities in your writing. Now, you notice a grammar problem that needs fixing. Now, a metaphor suggests itself. Now, you hear the voice of a workshop buddy reminding you that powerful essays often end the way they began.

---

152. Zinsser, *On Writing Well*, 251.

So, Elizabeth, thanks for the comment. Don't worry about keeping all the rules in mind. Keep writing. Keep learning. Keep looking at the words before you. The more you look, the more you see.

What Brand R U?

Writing ought either to be the manufacture of stories for which there is a market demand—a business as safe and commendable as making soap or breakfast foods—or it should be an art, which is always a search for something for which there is no market demand, something new and untried, where the values are intrinsic and have nothing to do with standardized values.

—Willa Cather, *Willa Cather on Writing*

If you write for a living—if you make a penny from your writing, or hope to—you have a brand. Maybe you have a logo, maybe you don't. Either way, you can't help but have a brand: a "name, term, design, symbol, or *any other feature* that identifies one seller's good or service as distinct from those of other sellers"[153] (emphasis mine).

You have a brand because no one else writes the way you do.

I can't believe I'm talking about personal branding. The term gives me the willies. I picture a sizzling iron rod, headed my way. I've never felt drawn to titles like *U R a Brand!* and *The Brand Called You.* But I can't argue with the searing power of association. Take a value-neutral image, like a green square. Now, give people good experiences whenever they find themselves near that green square, and watch them gravitate toward it. Or jab people with a stick whenever they look at that green square, and watch them avert their eyes.

---

153. "Brand," *American Marketing Association Dictionary,* accessed February 22, 2012, http://www.marketingpower.com/_layouts/Dictionary.aspx?dLetter=B.

Marketing departments spend billions creating associations with green squares. Coke is a green square. "Where's the beef?" is a green square. The Nike Swoosh is a green square. The Cornell University red square is a green square.

We are all, alas, green squares. For those of us who write professionally, our work does the marketing for us.

I have worked with a few writers who don't seem to get this, or to care. They avoid revising. For that matter, they avoid writing. They say, "It's close enough," or "People will know what I mean," or "Customers will figure it out." And I have to believe that those unlucky customers do figure it out—they figure out that they've been left to figure it out. Who hasn't *been* that customer, struggling to make sense of slapped-together assembly instructions, or unhelpful help topics? Who hasn't felt jabbed with the stick of careless writing? Who hasn't caught a glimpse of lackluster four-cornered greenness behind such writing and pledged (if only it were possible) to avoid further encounters?

How much better for a writer to ask, *Will customers figure this out?*—and then do what it takes to make sure the answer is yes. Let that be your brand. Or choose something else. Whatever fires you up, take a stand for it. Make your writing *your* writing. Explore ways to articulate your brand to yourself—to create your own green square—and then share that brand with the world through what you say and do.

You'll know that you've built a brand (or maybe multiple brands) when people say about you, "Oh yes, I know James. He's the guy who _____." Unless your name is Susan.

Consider the following ideas:

Write a Mission Statement

My mission statement applies broadly not just to my writing career but to all aspects of my life: grow continually and help others do the same. You may want your statement to focus on commerce, maybe zeroing in on one market (like healthcare writing) or a specialized skill (like XML authoring or humor). The form of mission statement you choose matters less than the thought you put into it.

The payoff comes when you face a big decision or set a major goal; your mission statement keeps you headed where you want to go. As a bonus, if you share your mission statement with others, they'll think of you when they hear about opportunities you'd want to know about— like a hospital down the road that's looking for an XML-authoring humorist.

Sum Up Your Writing Strengths and Interests in One Phrase

Make sure your phrase conveys a benefit to hiring organizations and that it differentiates you from others who apply for the same kind of jobs. I sometimes use the phrase *detail-oriented technical writer*. Put your phrase at the top of your résumé, business card, LinkedIn page, e-mail signature, or other professional profiles.

Prepare an Elevator Speech

When you find yourself in an elevator with your next company's CEO, who turns to you and asks, "What do you do?" you want to deliver a killer answer before the elevator reaches the next floor. My elevator speech used to be, "You know your VCR manual, the one that makes you feel so stupid that you've given up on recording movies? I write the other kind of manual." Of course, no one knows what a VCR is these days. Guess I need to follow my own next piece of advice.

Keep Your Elevator Speech Up to Date

How about this: "You know those boring books on writing that your English teachers subjected you to? I write the other kind."

Volunteer

You might not see yourself as a joiner. All the more reason to join something. A brand does you no good if no one knows about it. Join a writers' group, or a local chapter of a business association, or some other bunch of people who share your interests. (If you can think of it, someone nearby has started a Meetup for it.) Volunteer for a role that uses your writing skills—and that lets people see what you can do.

Got an elevator speech that won't quit? Turn it into a real speech, and present it at a meeting. Take great notes? Sign up for secretary. Did you solve a problem on the job recently? Turn your solution into a newsletter article. Born to persuade? Offer to run a membership drive or create killer promotional materials.

My involvement in groups like this has helped me find writing jobs—and fill them. I've made friends, seen fascinating demos, gained skills, and discovered excellent places to eat. Bust out of your comfort zone. Volunteer. See how it pays.

Blog about Something Important to You

If you don't want to start a blog, find a blogger you like and ask about contributing a guest post. Bloggers want well-written content that appeals to their readers, and some of them appreciate breaks from having to produce it all alone.

Make a Hash(tag) of Your Tweets

A hashtag—a # symbol plus a text string, like #ThisIsAHashtag—is a powerful symbol for getting your words seen. Even if you've never in your life sent a tweet (a brief message on Twitter), someday you may find it handy to know that hashtags give tweets visibility. Add a hashtag to any tweet, and you stand to reach a crowd.

Here's the tweet I posted about this essay when I first published it, in a slightly different form, on my blog:

What #Brand R U as a #writer? wp.me/1eWPK-IO #branding #writing #grammar #bloggers #techcom #xml #writetip #amwriting #writers

Anyone following any of these hashtags at the time could have seen this tweet. One follower of the #xml hashtag saw my essay and pointed to it in his e-newsletter, *The #XML Daily*.[154] What a thrill!

---

154. *The #XML Daily*, Mike McNamara, ed., Pubfluence, http://paper.li/aramanc /1329738333.

You can use hashtags (with or without capitals) with Twitter, Google+, and, probably, other social-media channels that I know nothing about. Hashtags come and hashtags go. No one controls them; people use them by unspoken agreement because they work. They give people a way to slip off into side rooms, away from the bustling party.

To find out which hashtags to use, observe. Search. Make up your own. Creative tweeter Aaron Gray says, "My fave use of hashtags is as metacommentary on the post itself. Yesterday, I used the tag #peopleplease and giggled."[155] You might discover in hashtags a new imaginative outlet. Even if only a few people see your most brilliant creations, those who do will #TakeNoteAndSmile.

> Even if only a few people see your most brilliant creations, those who do will #TakeNoteAndSmile.

Caution: Don't spam the universe with hashtag-heavy messages, though. As Gray warns, "One person's humorous metacommentary is another person's hashtag pollution."[156] Consider these delightful messages, which turned up when I searched to see what people were doing with the hashtag #Hashtag.

#Not #Everything #Needs #A #Hashtag #Remember #That.

You don't have to add #HashTags to Every tweet you Tweet
C'Mon Grow Up it looks tacky

In general, use social media wisely—not as a megaphone, but as a tool for conversation and listening—if you want it to help you build your brand. In the words of social-marketing guru Gary Vaynerchuk, "Brands and businesses [must] learn how to properly and authentically use social media to develop one-on-one relationships with their

---

155. Aaron Gray, Twitter @reply to the author, February 26, 2012.

156. Aaron Gray, comment on "Make a Hash(tag) of Your Tweets," *Word Power* blog, April 1, 2012, migrated to http://howtowriteeverything.com/make-a-hashtag-of-your-tweets.

customer base... Businesses that aren't able or willing to join the conversation will likely see their balance sheets suffer."[157]
#TakeThatToTheBank.

Start a Funniest Typos Bulletin Board at Work
Go low-tech. Analog. I'm talking corkboard. Tack up clippings or photos of amusing typos and grammar errors you've spotted, like *Support Our Scolarship Fund*. Raise awareness as you raise spirits. Colleagues' examples not allowed. Rating system optional.

Start a Radio Show or Podcast Series, or Get on Someone Else's
Are you the first in line on open-mic night? Got a thing for the spotlight? Create your own radio show, or volunteer as a guest on someone else's. For inspiration, tune in to Martha Barnette and Grant Barrett on *A Way with Words*, "public radio's lively language show,"[158] or Kristina Halvorson's series of podcast interviews, *Content Talks*.[159]

Write a Book
What better way to get your name out there than to put it on the cover of a book? Anyone can self-publish now. If you've got a book in you, go for it.

Gather a good team: readers—lots of readers—and an editor, a page designer, an illustrator, a publishing expert, a publicist. (You can do it all yourself, but you'll miss opportunities, and you might do something you'll regret, like mistyping Shakespeare's birth date.) If you hire professionals, prepare to spend a surprising amount of money. Consider it an investment in learning, like taking a course in publishing, except that instead of getting a grade, you end up with a sellable product.

Figure out how much time you'll need to pull your book together. Double that estimate. Then add a year.

Be willing to be wrong.

---

157. Gary Vaynerchuk, *The Thank You Economy* (New York: HarperCollins, 2011), 5, 22.
158. *A Way with Words*, http://www.waywordradio.org.
159. *Content Talks*, http://5by5.tv/contenttalks.

Venture into Merchandising

Selling stuff isn't for everyone, but some writers do it to build a brand, and they get some kicks along the way. Martha Brockenbrough, founder of National Grammar Day (March 4) and of the Society for the Promotion of Good Grammar (SPOGG), peddles proof-reader-pleasing mugs and T-shirts at "Shop SPOGG."[160]

> Figure out how much time you'll need to pull your book together. Double that estimate. Then add a year.

Grammar Girl Mignon Fogarty has two online shops, where she sells T-shirts, holidays cards, bags, and mouse pads that sport lines like *I've got a preposition for you, Squiggly's head is about to literally explode, To infinitives and beyond,* and *Talk grammar to me, baby.*[161]

Invent a Drink

Invite some friends over to celebrate National Grammar Day, or host a birthday party for the Bard. Mark the occasion with a drink of your own concocting: the Shakespeare Shooter or the Comma Kamikaze. Brock-enbrough calls hers the Grammartini.[162]

Be Green and Square

You're stuck with being a green square, so be green. Be square. Be proud.

Anyone for an XMLonball Splash?

---

160. Shop SPOGG, http://www.cafepress.com/spogg. (Don't miss this SPOGG logo created with latte foam by my favorite barista: "A Mug Shot We Love," March 29, 2012, http://grammatically.blogspot.com/2012/03/mug-shot-we-love.html.)

161. Behind the Grammar (http://behindthegrammar.com/shirts) and Quick and Dirty Tips Online Shop (http://www.cafepress.com/qdtshop).

162. For the Grammartini recipe and all kinds of other grammar-related goodies, see "National Grammar Day," *Grammar Girl* blog, accessed June 25, 2012, http://nationalgrammarday.com.

Postscript

The blog version of this chapter invited people to submit XMLonball Splash recipes.[163] My only taker, Scott Abel, submitted a witty recipe, about which I then posted a tweet (with hashtags), which caught the eye of Leisa LaDell, who promoted the story in her e-newsletter, shown here. See that little green square peeking out from behind Scott's picture?

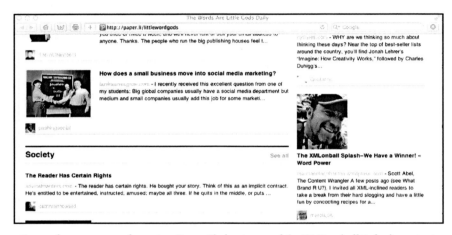

E-newsletter excerpt featuring Scott Abel, winner of the XMLonball Splash contest.

---

163. Marcia Riefer Johnston, "Recipe Contest: The XMLonball Splash—We Have a Winner!" *Word Power* blog, May 13, 2012, migrated to http://howtowriteeverything .com/recipe-contest-the-xmlonball-splash-we-have-a-winner.

Who's Your Sam?

It is human nature to imagine, to put yourself in another's shoes.
—GERALDINE BROOKS, "TIMELESS TACT HELPS SUSTAIN A LITERARY TIME TRAVELER," *NEW YORK TIMES*

As I meandered through a confusingly laid out hospital, which shall remain nameless, this sign caught my eye. I say *caught my eye* advisedly. You'll soon understand why.

First, I struggled to figure out what the information below the line meant. (My car was in Parking 2. Should I use this door?) After I solved that mystery—yes, this door would take me to my car—I puzzled over the presence of Braille. Someone had spent money and effort to (theoretically) share this information with the sightless. But for whom, exactly, could these tactile words have value?

I did my best to imagine such a person. Flipping a coin, I made this person male. I'll call him Sam. What do we know about Sam?

- Sam is blind.

- Sam can read Braille.

- Sam knows that the sign exists. (Otherwise, he would never feel for it.)

- Sam is not in a wheelchair. (The sign is five feet above the floor. If Sam were in a wheelchair, he'd have to stretch to find it.)

- Sam has a companion who's in a wheelchair. (The message below the line has value only to people looking for a wheelchair-accessible route to Parking 2. If no one's in a wheelchair, no one cares about the message.)

- Sam's companion is unable to read the sign aloud. (Why else would Sam take the time to read with his fingers?) Ergo, his companion must be blind too, or else illiterate, or mute, or too young to read. I picture a five-year-old. I'll call her Julie.

- Sam is unfamiliar with the wheelchair-accessible routes, and he has no companion capable of guiding him. (If he knew the routes or had a guide, he'd have no use for the sign.)

- Sam wants to get to Parking 2. (The sign exists solely to direct people to Parking 2.)

Who is Julie to Sam? A neighbor girl? His niece? No way. He'd never be sent out alone to take her to the hospital. She has to be his daughter. He could be a single dad, proud and capable. Maybe Julie fell and broke her leg. Maybe the two of them recently moved to Portland (now you know) and have no nearby friends or family.

We come now to the scenario: Sam and Julie found their way into the hospital. (We have to go with that.) They completed their business there and are ready to go home. Sam is now wheeling Julie down a hospital corridor with one hand, feeling his way along the wall with the other.

"Hang in there, Punkin. I'm sure there's a sign here somewhere … Aha!"

Sam's fingers glide along the first row of dots: *Parking 2 Level G.*

"Parking 2 Level G. That's what we're looking for."

Stairs to Level G.

"Hmm. We can't take the wheelchair down the stairs."

Not an Accessible Route.

"That must mean we can't take the wheelchair down the stairs. Wow. Thanks for that."

Use Ramp to Parking 2 Level H.

"Okay. Where's the ramp?"

[_____]

"What the ...?"

Hell-o-o! Sam doesn't need a ramp to the parking area. Sam can't see to drive a car. Sam's never going to feel his way along these walls while single-handedly wheeling his little girl through a labyrinth.

Whoever requisitioned this Braille translation had no conceivable audience in mind. However well-intentioned, that person (or that committee or that body of regulators) didn't think through the scenario, didn't play the whole movie.

> **Whoever requisitioned this Braille translation had no conceivable audience in mind.**

How about you? What do you see happening on the receiving end of your messages? How fully do you imagine your readers? Can you hear them talking, see them strolling down the halls? Do you play the whole movie? Who is your Sam?

Mastering the Art of Knowing Your Audience

The best time for planning a book is while you're doing the dishes.
—Attributed to Agatha Christie

If the editor-in-chief at Houghton Mifflin had had his way in 1959, Julia Child's *Mastering the Art of French Cooking*—a bestseller and arguably the most influential work in American cookbook history—would never have been published. Even after Julia spent a year tightening her manuscript at this editor's request, condensing it as far as she felt she could, the editor rejected it, judging it "so huge, expensive, and elaborate that it was certain to seem formidable 'to the American housewife.'"[164]

This editor thought he knew Julia's audience. He thought her capacious creation would send readers straight to Snoozeville. Oops.

Julia knew her audience. Seeing that Houghton Mifflin did not, she changed publishers. In 1961, Alfred A. Knopf served up this monumental *pièce de résistance*—all 726 pages of it—to the cooking world. That world, like Knopf's, has never been the same.

Know your audience. It's Rule #1. It's simple. It's fundamental. And it's as easy to dismiss as a check box. (Writers always know *something* about their audience.) This rule skirts the question that matters: how well should I know my audience? After reading Laura Shapiro's *Julia Child: A Life*, I'm more convinced than ever that if I want my words to be read—whether I'm writing for pleasure or for pay—I must envision my readers, as Julia did, precisely and constantly.

---

164. Laura Shapiro, *Julia Child: A Life* (New York: Penguin, 2009), 86.

Julia's envisioned reader was a virtual companion, an inseparable tagalong, full of curiosity. This persona, a composite of "young bride" and "chef-hostess," focused and energized Julia's writing, as Shapiro describes at length:

> This imagined reader…who couldn't cook until the right book fell into her hands, had a permanent place in Julia's consciousness and directly inspired the immense amount of detail that characterized her recipes. Like a ghost from Julia's own past, she trailed Julia from kitchen to desk and back again, forever trying to figure out whether the roast was done, why the chops were steaming in the pan instead of browning properly, what made the cream puffs soggy, and exactly how thick the beef slices should be: a quarter inch? an eighth of an inch? [165]

Julia wrote specifically for people interested in creating peak experiences with food. These readers were so real to her, and so important to her, that she began her foreword in *Mastering* by describing them:

> This is a book for the servantless American cook who can be unconcerned on occasion with budgets, waistlines, time schedules, children's meals, the parent-chauffeur-den-mother syndrome, or anything else which might interfere with the enjoyment of producing something wonderful to eat. [166]

If only I could always see my readers this clearly!

Too often, I have only a superficial sense of the human beings on the receiving end of my efforts. For example, when I'm hired to describe how to install or use or repair some device, it's often difficult to get enough information about the installers/users/repairers to enable me to make solid decisions on their behalf. I don't know how familiar they are with this product or with similar products. I don't know what terms and concepts they already understand. I don't know their surroundings, motivations, frustrations, or burning questions.

---

165. Ibid., 65–66, 96.

166. Julia Child, Louisette Bertholle, and Simone Beck, *Mastering the Art of French Cooking* (New York: Knopf, 1961), vii.

Such indistinct personas—unlike Julia's easy-to-conjure, would-be gourmets—can't tell me what they need. I can't hear them. I can't see them. When I don't know my audience, even if I write well, I might not write well *for them*.

Writers rarely have the kind of ideal relationship with their audiences that Julia had with hers. As technical communicator Mark Baker notes on my blog, Julia "knew her audience because she was her audience."[167] In writing for herself, she created an audience

> *Know your audience.*
> **It's Rule #1. It's simple.**
> **It's fundamental. And it's as**
> **easy to dismiss as a check box.**

like herself. Most of us don't get to do that. We have to cross a gap. The bigger the gap between us and our readers, the harder we have to work to understand what they need.

How do we come to understand what our audience needs? Well-researched, well-crafted personas help. Like characters in a transporting piece of fiction, personas bring us into someone else's world. They help us decide what to leave out, what to add, what words to use, and many other particulars. When we write for people whose background or interests differ from our own—especially if we work with a group that's supposed to pull together to create something appealing for that audience—we, and the organizations we work for, can profit from personas.

Businesses profit literally: the strategic use of personas brings in money.[168] Writers who work with personas in mind may make money too. For sure, we profit in terms of satisfaction. At least I find it satisfying to write for a reader whom I can imagine fully and accurately.

---

167. Mark Baker, comment on "Mastering the Art of Knowing Your Audience," *Word Power* blog, September 16, 2011, migrated to http://howtowriteeverything.com /know-your-audience-personas-readers.

168. Recommended reading on personas:
- Janice (Ginny) Redish, "People! People! People!" in *Letting Go of the Words: Writing Web Content that Works* (San Francisco: Morgan Kaufmann, 2007), 11–28.
- Alan Cooper, "Designing for Pleasure" in *The Inmates Are Running the Asylum: Why High-Tech Products Drive Us Crazy and How To Restore the Sanity* (Indianapolis: Sams, 2004), 123–147.
- John Pruitt and Tamara Adlin, *The Persona Lifecycle: Keeping People in Mind Throughout Product Design* (San Francisco: Morgan Kaufmann, 2006).

Writing for someone I know—someone real or imagined, someone just like me in many ways or in few—is like making a gift for a friend. While I'm working on it, whatever it is, I confidently imagine the recipient opening it and saying, "Yes!" Julia must have felt this same confidence every time she placed her book in the hands of an eager young cook.

Decisions, Decisions

> *To choose wrongly is to leave the hearer or reader with a fuzzy or mistaken impression. To choose well is to give both illumination and delight.*
> —S. I. Hayakawa, *Choose the Right Word*

William Zinsser's classic book *On Writing Well* includes a chapter called "A Writer's Decisions" in which he "dissects" one of his own articles, explaining some of the "countless successive decisions" that he made as he wrote it.[169] Since I found the peek into his decision-making process enriching, I offer here a peek into my own. Like Zinsser, I analyze an essay that I wrote on returning from a trip. Even if you don't write narrative nonfiction, I hope that as you read this essay and analysis, you'll find yourself thinking, *Aha! I see what's at work there, and there, and there. I see how I could use that principle myself.*

I chose this essay because it demonstrates most of the messages delivered in this book, including these: write about what you know, write about what you love, and put what's most important last.

The Essay

Coming to Terms: My Children and I Enter Foreign Territory
  Walking down the narrow sidewalk ahead of me, my son, nearly out of his teens, elbows my daughter, newly out of hers, and points to the Volcán de Agua, which has popped through the clouds as she's been promising us it would.

---

169. Zinsser, *On Writing Well*, 245.

"Look at that! Would you look at that volcano? Just look at it."

He's channeling a loopy Ed Bassmaster character. I know this because the instant we walked into the Posada Hotel Burkhard—"Wi-Fi! Yes!"—he fired up his laptop to show his sister and me his favorite Bassmaster videos on YouTube.

"Your *mom's* a volcano," Elizabeth says.

"*Your* mom's a volcano," Brian says.

I say, "Your mom's your *mom*," an in-joke that dates back to a moment of silliness years earlier, a line that any of us might say at any time.

Brian and I follow Elizabeth's flip-flops down the wet cobblestone streets of Antigua, where we will spend the next two days. Guatemalans call this time of year (it's July, the middle of the rainy season) *invierno*, winter, as if they were south of the equator or as if they needed parkas instead of occasional sweaters. This usage resonates with me—its upside-downness. Upsidedown is how I feel in this place where I have no bearings and can neither navigate nor communicate on my own.

A year has passed since Elizabeth flew off, freshly graduated from college, to start her Peace Corps service. Even before she left the States, she started working on her brother and me to plan a visit. Neither of us hankered to see Central America. She persisted. I relented. Eventually, Brian came around. We got the passports, the shots, the airplane tickets, the raincoats, the translation apps. A few days after Brian graduated from high school, we boarded a plane in Syracuse. We flew over the Finger Lakes area, where I had spent most of my adult years, where my children had grown up, where winter was winter. *Click!* went Brian's camera.

Now that we've arrived, I relax. No more planning, anticipating what-ifs. I give myself over to the moment. The rain has stopped. The Volcán passes from view as the three of us turn a corner. Sunlight welcomes us into the lush, wide-open block known as Parque Central. The sidewalk leads to a fountain. Two sparkling arcs stream from a mermaid's breasts, which she placidly cups as if to say, "Look at them. Would you just look at them?" *Click!* goes Brian's camera.

All around the park, grime and decay are defied by the strident colors that shout from ancient buildings, tropical flowers, *traje*-clad women, baskets full of painted trinkets, and piles of embroidered cloth goods. Electric greens, oranges, pinks, scarlets, blues, whites, purples, and yellows zing along my optic nerve to that pleasure spot just this side of pain. I can't help saying over and over, "Would you look at that?"

What a rush, sharing this jarringly rich experience with these two young people whom I brought into the world and who have so recently done what once seemed impossible: crossed over into adulthood. How freeing to realize that for this couple of weeks—maybe forever—they will need me less as a parent than as a companion.

In fact, on this trip I am mostly a follower, one who struggles to keep up. I don't mean physically. *Al contrario*, at Copán I rarely lose sight of Brian and Elizabeth as we all scramble up and down the rough, moss-covered blocks of the eerie Mayan ruins. Ditto Lake Atitlán, where we climb the seemingly endless stone steps up the hillside to our quarters. No, my struggle is mental. I'm talking about keeping up with the conversation.

We're sitting at Frida's Cocina Mexicana, drinking margaritas (which nineteen-year-old Brian can do here) and eating tacos. Elizabeth's recently bobbed brown curls stop just short of her shoulders. They hang loose and soft around her face as if to say, "What's not to like about these curls?" When she was little, she didn't want her hair short because people took her for a boy; for years she straightened it or tied it back. How far she has come. And here sits Brian, relaxed in his slim, chiseled wrestler's body, his hair the same straight, dark blond that mine used to be. When he was little, he said he'd never take any trip that required him to get shots. How far he, too, has come.

A familiar song is playing.

Brian (challenging): "Who's that?"

Elizabeth (instantly): "Adele. Which album?"

Brian (after a few seconds, deflated): "Your mom's."

We order another round of tacos. We talk. In someone else's voice, seemingly out of nowhere, Brian warns: "I don't want no scrubs." When our blank looks tell him that his audience doesn't get the reference, the unappreciated artist wags his head and says, "You have to see *The Other Guys*." Later, I'll Google the movie, and I'll hear the quirky delivery of the scrubs line. I'll see that Brian nailed it. I'll discover that the character has, in turn, borrowed the line from a TLC song, making Brian's quote a quote within a quote.

My children are otters frolicking in the stream of pop culture.

Occasionally, I score a point. Back at our hotel, we're playing Monopoly. Elizabeth, the banker, has just sold her brother a handful of little green houses. He spills them onto Boardwalk and Park Place.

Brian: "Zoop!"

Elizabeth: "I put you in this game. I'll take you out."

Mom: "Bill Cosby!"

Elizabeth: "Mom got it." (I don't deserve the credit—they got Cosby from me—but, hey.)

Mom: "Zoop!"

This staccato, falsetto syllable punctuates our Guatemala days. It took me a few repetitions to figure out the snappy little sound. It indicates connection, things clicking into place. The origin is an episode of *Family Guy*. (Brian played this video for us too.) A cartoon alien opens its mouth and shoots out a littler mouth that lives on the end of its tongue. The mouths exchange words. Then the big mouth retracts the little one—Zoop!—pulling its own back into itself.

I can't pull my little ones back. Elizabeth has returned to western Guatemala, where she's bringing computers to schoolchildren. Brian is starting college in western New York. I'm settling in to a new life on the West Coast, thousands of miles from either of them. For the first time since they were born, I have no idea when or where I will see my children next.

I will get used to this phase: the separation, the uncertainty. For now, I pore over my vacation photos. I'd love to show them to you. Of course you'll never see Brian and Elizabeth as I do. But oh, what splendid human beings! I mean, look at them. Would you just look at them?

The Analysis

Let's look first at audience and purpose, the foundational decisions of any writing effort. I had in mind a broad audience and a lofty purpose. I wanted to appeal to anyone who has experienced the simultaneous forward and backward tugs of a major life transition. I wanted to transport those readers, to capture a moment in a way that would resonate. I wanted to do for them what my favorite writers do for me: heighten their sense of awe at being alive.

Next decision: genre. What type of piece would do what I hoped to do? I chose the personal essay for the same reason that Jack Hart chooses this genre: "I usually turn to personal essay when I've been emotionally affected by something I've experienced, without quite knowing why."[170]

As for organization scheme, I hardly needed to think about how to structure this piece. Information about a Guatemala

> **I wanted to do for readers what my favorite writers do for me: heighten their sense of awe at being alive.**

trip could theoretically be organized alphabetically ("A is for avocado trees, which grow in every yard...") or geographically ("First, we hit the Mayan ruins in the southeast...") or procedurally ("When planning a trip to Central America, follow these ten easy steps..."). But I had a story to tell, and I wanted to send it straight to the heart. What could I choose but narrative: someone does this, someone does that, flashback, something happens, turning point, someone realizes something, the end.

Two more high-level decisions made themselves without debate: tone (intimate, relaxed) and diction level (informal, colloquial). I used everyday words (*loopy*), contractions (*We've arrived*), phrasal verbs (*came around*), and other slang (*nailed it*) as if I were writing a letter to a friend.

For the rest of this analysis, let's move through the essay in sections.

---

170. Hart, *Storycraft*, 209.

Title

Coming to Terms: My Children and I Enter Foreign Territory

This title had many predecessors—discarded versions that said too much or too little, or that lacked play, or that simply didn't feel right. A title often takes its time arriving. A title must accomplish two things: flag readers down and convey (or at least hint at) the piece's main theme. Ideally, the sound and rhythm also hit the ear well, and the wording evokes multiple meanings.

Identifying a piece's main theme can require a surprising amount of thought. Writing coach Jack Hart urges writers (writers of narrative nonfiction at least) to hold off on choosing a title until they have created a one-sentence "theme statement"—specifically, a *noun–transitive verb–noun* statement, like "Frugality causes waste" or, in the case of his own book, "Stories wring meaning out of life." In Hart's experience, which includes guiding several Pulitzer Prize–winning narratives to publication, this type of theme statement acts as "a kind of elixir" and "helps you find a title." If you're writing a nonfiction piece, why not apply this elixir yourself? Before you entertain any candidates for your title, capture your theme in a sentence: "one clear, coherent sentence that expresses a story's irreducible meaning."[171]

Here is my essay's theme statement: life whisks us ever onward.

I picked my title for several reasons. For starters, the phrase *coming to terms* plays on the word *terms*, as in the Spanish terms that my son and I kept bumping into, not to mention the English terms that my lingo-savvy children kept challenging me with. *Coming to terms* also applies to the main theme of coming to terms with a new phase of life. The phrase *foreign territory*, which refers, of course, to the faraway country where my children and I spent those two weeks together, also applies to the new relationship—adult to adult—that the three of us were entering into.

---

171. Hart, *Storycraft*, 143–145. In the statement about "a story's irreducible meaning," Hart is quoting "guru of narrative" Robert McKee. Hart notes that a theme statement helps writers and editors find not only a story's title but also its "shape."

Let's turn to the body of the essay, one excerpt at a time.

> **Excerpt 1**
>
> Walking down the narrow sidewalk ahead of me, my son, nearly out of his teens, elbows my daughter, newly out of hers, and points to the Volcán de Agua, which has poked through the clouds as she's been promising us it would.
>
> "Look at that! Would you look at that volcano? Just look at it."
>
> He's channeling a loopy Ed Bassmaster character. I know this because the instant we walked into the Posada Hotel Burkhard—"Wi-Fi! Yes!"—he fired up his laptop to show his sister and me his favorite Bassmaster videos on YouTube.
>
> "Your *mom's* a volcano," Elizabeth says.
>
> "*Your* mom's a volcano," Brian says.
>
> I say, "Your mom's your *mom*," an in-joke that dates back to a moment of silliness years earlier, a line that any of us might say at any time.

Several types of decisions went into writing this opening. For example, for thousands of years, storytellers have started their stories *in medias res*—in the middle of things. Homer's *Iliad* opens in the middle of the Trojan War. Similarly, the *Odyssey*, that colossal story of stories, opens seven years into Odysseus's journey home. My own essay, a reflection on a shorter odyssey, follows in this tradition of ensnaring people with a scene-in-progress rather than explaining right off who we are, where we are, or how we got there. You can open *in medias res* in almost any kind of narrative, fiction or nonfiction. If you aren't telling a story—if you have no *res* to jump into the *medias* of—you still want to figure out some way to grab readers quickly.

You could grab readers with a quotation. Spoken words, like Brian's Bassmasterisms, impart information—and draw readers—like nothing else. You could slap quotation marks around any old text, but the best quotations capture a distinct turn of phrase, nail an idiom, convey the voice of an unmistakable speaker. As the song goes, "It don't mean a thing if it ain't got that swing."

We might as well discuss right up front a type of decision that writers face at every writing moment: word choice. When you consider all that goes into choosing even one word, you have to wonder how any writing gets done. Take the word *loopy* in "He's channeling a loopy Ed Bassmaster character." How did I choose this word? Slowed way down, the process goes like this:

- Does this word imply anything that I don't mean? (From several sources, I confirm that *loopy* means exactly what I have in mind: "crazy" or "silly." Its other meanings won't confuse or mislead anyone.)

- Could any other word convey my meaning more accurately or more fully? (No better word comes to mind. *Goofy* could work, but it implies a foolishness that doesn't fit. I want a word that implies simply "a bit off." A synonym search brings up nothing that beats *loopy*.)

- Does this word have the right tone and diction level? (Yep. It's informal and colloquial. I would say *loopy* in a letter to a friend.)

- Have I overused this word? (Oh, the beauty of electronic searching. Nope. *Loopy* appears only once in my doc.)

- How about its sound and rhythm—anything good happening there? (Yes! *Loopy* makes me think of Brian's exaggerated gait and sing-song voice when he's "doing" Bassmaster. I can hear him now: "Well, *loopy, loopy, loopy*. That's some fine syllables there. Would you just listen to them!")

Ladies and gents, we have chosen a word.

One more thing about Excerpt 1: it makes me laugh. I don't know how to be funny, but I like to share things that I find funny. If you want to hit the reader's funny bone, write with integrity. Convey what you find funny as truly as you can, leaving unstated whatever people will get on their own. The most satisfying laughter comes from recognition.

> **Excerpt 2**
>
> Brian and I follow Elizabeth's flip-flops down the wet cobble-stone streets of Antigua, where we will spend the next two days. Guatemalans call this time of year (it's July, the middle of the rainy season) *invierno*, winter, as if they were south of the equator or as if they needed parkas instead of occasional sweaters. This usage resonates with me—its upside-downness. Upside-down is how I feel in this place where I have no bearings and can neither navigate nor communicate on my own.

Excerpt 2 brings to mind other decisions. For one thing, even though I wanted to communicate something abstract (awe, aliveness), I focused, here, on the concrete: flip-flops on wet stone streets. The way to readers' hearts is through their senses. Readers must see it, smell it, hear it—feel it under their feet—before they can care about it.

I introduced our location surreptitiously in Excerpt 2. Rather than announce, "Hey, folks, by the way, we are in Antigua, Guatemala," I slipped in "streets of Antigua," and I mentioned that Guatemalans call this time of year *invierno*. You know where we are even though I didn't tell you straight out. When you write, the less noticeably you weave in the exposition (background information) the more engaged the reader stays.

The final sentence in Excerpt 2 ("Upside-down is how I feel...") demonstrates another technique that aids reading: transition by repetition. In general, transitions pass along the baton of thought from one sentence to another, one paragraph to another, one section to another. Standard transitions include phrases like *in the first place* and *on the other hand*. Transitions don't have to take a standard form, though. Some of the most effective transitions simply repeat key words from previous sentences, as in the repetition of *upside-down* here: "This usage resonates with me—its upside-downness. Upside-down is how I feel in this place." (Repetition of this type—words from the end of one sentence or clause repeated at the beginning of the next—is an example of the rhetorical device known as anadiplosis.)

My decision to put this transition word (*upside-down*) at the beginning of the second sentence involved a couple of mini-decisions. The second sentence could have read this way instead: "I feel upside-down in this place ..." But I wanted a more graceful progression, a smoother handoff. As if maneuvering one party guest toward another for an introduction, I moved *upside-down* to the beginning of its sentence, next to its predecessor, *upside-downness*. To do this, I had to use *is*: "Upside-down is how I feel ..." I squinted at this trade-off, since I avoid *be*-verbs (see "*To Be* or Not *To Be*" on page 13). I decided to live with this *is* because bumping *upside-down* to the beginning accomplished two things. First, it connected the two sentences instantly, as if to say, "Speaking of upside-downness ..." Secondly, it emphasized the key word *upside-down*. (Strunk and White call the beginning "the other prominent position in the sentence."[172])

Writing decisions that involve conflicting guidelines (like smooth transition vs. *be*-verb) pop up all the time. In these cases, ask not, *What's the right thing to do?* but *What's the best thing to do here?*

Next in the narrative comes a flashback. Anyone who has read this far wonders, *Who are these people? Where are they from?*

Excerpt 3

A year has passed since Elizabeth flew off, freshly graduated from college, to start her Peace Corps service. Even before she left the States, she started working on her brother and me to plan a visit. Neither of us hankered to see Central America. She persisted. I relented. Eventually, Brian came around too. We got the passports, the shots, the airplane tickets, the raincoats, the translation apps. A few days after Brian graduated from high school, we boarded a plane in Syracuse. We flew over the Finger Lakes area, where I had spent most of my adult years, where my children had grown up, where winter was winter. *Click!* went Brian's camera.

---

172. Strunk and White, *Elements of Style*, 2ⁿᵈ ed., 26 (page 53 in the illustrated 3ʳᵈ edition).

Many decisions came into play in Excerpt 3, including when and how to shift tenses from present to past to earlier past and then back to present. It takes effort to stitch time frames together so that the seams don't show. Verb-tense shifts proved surprisingly challenging throughout this essay (as they are proving throughout this analysis: "I chose..." in past tense vs. "Hemingway warns..." in literary present tense, for example). For the record, any verb-tense inconsistencies you find in my essay (or in this analysis) result—I mean resulted—from careful deliberation...or from confusion, I can never tell which. If your writing requires frequent shifts in tense, you'd do well to read up on the subtleties involved in this type of decision making.[173]

In the middle of Excerpt 3, we come to a pair of two-worders: "She persisted. I relented." Several writerly things are going on here. First, these sentences are dramatically shorter than those around them. Spotlight! In addition to their unifyingly unique stubbiness, their similar structure—subject verb, subject verb—reinforces their pairing and their significance. (The type of parallelism shown here at this pivot point—parallel grammatical structure enhanced by the tension of logically spring-loaded contrasting ideas, namely, persisting and relenting—is an example of the rhetorical device known as antithesis.)

The end of Excerpt 3 swings us back into action: "*Click!* went Brian's camera." The narrative, although still in past tense, is moving forward again. This *Click!* also awakens readers' ears. It returns them from the world of memory to the world of the senses.

> **Excerpt 4**
>
> Now that we've arrived, I relax. No more planning, anticipating what-ifs. I give myself over to the moment. The rain has stopped. The Volcán passes from view as the three of us turn a corner. Sunlight welcomes us into the lush, wide-open block known as Parque Central. The sidewalk leads to a fountain. Two sparkling

---

173. See, for example, Plotnik, "Tense: A Sticky Choice," in *Spunk & Bite*, 44–52.

> arcs stream from a mermaid's breasts, which she placidly cups as if to say, "Look at them. Would you just look at them?" *Click!* goes Brian's camera.

We're back in the moment, present tense. What can I say about Excerpt 4 that I haven't already said? If a motif—an element repeated throughout a piece, like the phrase *Would you look at that?*—presents itself as you develop your draft, use it artfully. Weave it through your writing as you would an accent color through a piece of fabric, allowing it to peek out just often enough for people to recognize it and take pleasure in it. Ideally, in the end, those spots of color achieve a balanced, unifying effect that resonates in a way that no one saw coming.

Reading Excerpt 4 for not only the content but also the language, you might linger on the phrase *sparkling arcs*. Notice the dum-dee-dum rhythm. The soft *ss* at the beginning and end. The onomatopoetic sparkle of *sparkling*. The *ark-ark* echo of *sparkling arcs*, a double hit of aural loveliness: consonance plus assonance. (Assonance—here, the *a-a*—is the repetition of vowel sounds to create internal rhyming. Consonance—here, the *rk-rk*—is the repetition of consonant sounds within words. Both assonance and consonance are kinds of alliteration in the broad sense of the term: the repetition of sounds. Used in small doses, alliteration "can put an innocent hop and skip" in your prose.[174]) This overall combination of sound effects, laid over a just-right meaning, takes me someplace that only beautiful language can. No phrase is inherently beautiful, of course. *Sparkling arcs* might do nothing for you. But it pleases me. I could say it over and over. I did, in fact, as I was writing this description. When you write, you must please yourself (ears and all) before you can hope to please anyone else.

Excerpt 4, like the rest of this essay, originally included more words. I constantly add and cut, add and cut. Then, when I think I'm done—sometimes with an editor's (eventually welcome) nudging—I cut some more. Here, for example, I deleted "with her hands." How else would a mermaid cup her breasts?

---

174. Plotnik, *Better than Great*, 219.

Finally, Excerpt 4 gives me a chance to make this point: action, well captured, speaks for itself. Consider the line "*Click!* goes Brian's camera." Readers don't need to be told "Brian *was* looking." Nor do readers need to be told that I smiled as he took that photo. They don't need to be told what I felt, or even that I experienced a number of emotions too complex and too fleeting to merit explication here. As Hemingway famously says,

> The greatest difficulty, aside from knowing truly what you really felt, rather than what you were supposed to feel … was to put down what really happened in action; what the actual things were which produced the emotion that you experienced. The real thing, the sequence of motion and fact which made the emotion … would be as valid in a year or in ten years or, with luck and if you stated it purely enough, always.[175]

Excerpt 5

All around the park, grime and decay are defied by the strident colors that shout from ancient buildings, tropical flowers, *traje*-clad women, baskets full of painted trinkets, and piles of embroidered cloth goods. Electric greens, oranges, pinks, scarlets, blues, whites, purples, and yellows zing along my optic nerve to that pleasure spot just this side of pain. I can't help saying over and over, "Would you look at that?"

The description of the park in Excerpt 5 opens with a rare instance of passive voice ("are defied") contributing value to a sentence. In this case, passive voice enables *grime and decay* to come first so the sentence can end with an optimistic trumpeting of all the colorful things that do the defying. Not only does the passive voice put the emphatic things—the beautifully colored things—at the end (a key principle in writing powerful sentences), but it also creates a sentence whose structure mimics its meaning: "Grime and decay, you are going *down!*" Passive voice doesn't last long, though. Active voice returns in the form of the onomatopoetic verb "zing."

---

175. Hemingway, *Death in the Afternoon*, 11–12.

What else is going on in Excerpt 5? Several things. The sentences contain the simplest of all types of parallelism: a daisy-chain of nouns (buildings, flowers, women) followed by a cascade of colors (greens, oranges, pinks). Also, this paragraph treats you to bits of cadence, as in the phrase *just this side of pain*, which practically invites you to nod along. And the motif "Would you look at that?" peeks out again, emphasized by its placement at the end of the paragraph—and by its utterance, for the first time, by me.

Excerpt 5 includes the Spanish word *traje*. We haven't yet touched on the handling of foreign words. In this essay, and in general, I follow in the tradition of stylists who avoid foreign words unless the context cries out for them. Even then, I sprinkle them in sparingly—like dots of Picamás sauce on a burrito, you could say—and I use context to help convey meanings. (Notice that you didn't need to be told just now that Picamás sauce is Guatemalan or that it's hot.) Wherever possible, define terms—even English terms that some readers might not know—through context. In Excerpt 5, when readers see *traje* joined to *clad*, they gather that *traje* refers to the traditional outfits of the region. And because the phrase *traje-clad women* belongs to a series of colorful things, readers don't need to be told that this *traje* shimmers like a box of crayons. The context says it all. When context can't say it all, translate with minimal disruption. In an earlier sentence—"Guatemalans call this time of year…*invierno*, winter, as if they were south of the equator"—the in-line translation for *invierno* makes hardly a ripple.

The next passage, shown in Excerpt 6, introduces the concept that my parental role is changing as my children move into adulthood.

> **Excerpt 6**
>
> What a rush, sharing this jarringly rich experience with these two young people whom I brought into the world and who have so recently done what once seemed impossible: crossed over into adulthood. How freeing to realize that for this couple of weeks—maybe forever—they will need me less as a parent than as a companion.

> In fact, on this trip I am mostly a follower, one who struggles to keep up. I don't mean physically. *Al contrario*, at Copán I rarely lose sight of Brian and Elizabeth as we all scramble up and down the rough, moss-covered blocks of the eerie Mayan ruins. Ditto Lake Atitlán, where we climb the seemingly endless stone steps up the hillside to our quarters. No, my struggle is mental. I'm talking about keeping up with the conversation.

Even as Excerpt 6 touches on the abstract notion of children growing up, it stays grounded in concrete language. The reader feels those rough, moss-covered blocks, imagines running out of breath…stays engaged.

The first sentence in Excerpt 6 includes the phrase *jarringly rich experience*. Adverbs, including *-ly* words, often add no value (as in, *Frankly, I'll tell you the truth*). Adverbs do add value, though, when they contribute surprise. Since *jarringly* and *rich* don't normally go together, their pairing conveys a meaning more complex than either word conveys alone, while it pleasantly prickles the ear.

The final sentence of Excerpt 6—"I'm talking about keeping up with the conversation"—pivots into a central scene, the restaurant.

Excerpt 7

> We're sitting at Frida's Cocina Mexicana, drinking margaritas (which nineteen-year-old Brian can do here) and eating tacos. Elizabeth's recently bobbed brown curls stop just short of her shoulders. They hang loose and soft around her face as if to say, "What's not to like about these curls?" When she was little, she didn't want her hair short because people took her for a boy; for years she straightened it or tied it back. How far she has come. And here sits Brian, relaxed in his slim, chiseled wrestler's body, his hair the same straight, dark blond that mine used to be. When he was little, he said he'd never take any trip that required him to get shots. How far he, too, has come.

Knowing that more conversation lies just ahead, readers can abide some action-stopping description here: hair color, etc. These details help them visualize the scene while deepening their understanding of the people being described (and of my perspective on them).

A punctuation note: In the sentence "When she was little, she didn't want her hair short because people took her for a boy; for years she straightened it or tied it back," I debated whether to insert a comma after "for years." Many style guides support omitting the comma when an introductory phrase has only two or three words. In this case, the comma would have added choppiness without adding value. I opted for the omission. Even punctuation decisions sometimes depend on context.

Next comes the awaited conversation.

Excerpt 8

A familiar song is playing.

> Brian (challenging): "Who's that?"
> Elizabeth (instantly): "Adele. Which album?"
> Brian (after a few seconds, deflated): "Your mom's."

We order another round of tacos. We talk. In someone else's voice, seemingly out of nowhere, Brian warns: "I don't want no scrubs." When our blank looks tell him that his audience doesn't get the reference, the unappreciated artist wags his head and says, "You have to see *The Other Guys*." Later, I'll Google the movie, and I'll hear the quirky delivery of the scrubs line. I'll see that Brian nailed it. I'll discover that the character has, in turn, borrowed the line from a TLC song, making Brian's quote a quote within a quote.

My children are otters frolicking in the stream of pop culture.

Occasionally, I score a point. Back at our hotel, we're playing Monopoly. Elizabeth, the banker, has just sold her brother a handful of little green houses. He spills them onto Boardwalk and Park Place.

> Brian: "Zoop!"
> Elizabeth: "I put you in this game. I'll take you out."
> Mom: "Bill Cosby!"
> Elizabeth: "Mom got it." (I don't deserve the credit—they got Cosby from me—but, hey.)
> Mom: "Zoop!"

Want people to read something? Put it in dialogue. When you write dialogue, you write for the eye and the ear at once. The eye can't resist those invitingly short lines. And we all like to listen in on conversations. Words that come out of someone's mouth tell us a lot—about what's being said and about the person or character saying it. When we hear (or imagine hearing) people talking, we're instantly drawn into a moment, an interaction unfolding. We decipher meaning from the speakers' word choices and from the context of the scene. We take mental notes on ways we might want to talk—or not talk. We imagine the feelings behind the words and the things left unsaid. We wonder, *What will they say next?*

Excerpt 9

This staccato, falsetto syllable punctuates our Guatemala days. It took me a few repetitions to figure out the snappy little sound. It indicates connection, things clicking into place. The origin is an episode of *Family Guy*. (Brian played this video for us too.) A cartoon alien opens its mouth and shoots out a littler mouth that lives on the end of its tongue. The mouths exchange words. Then the big mouth retracts the little one—Zoop!—pulling its own back into itself.

I can't pull my little ones back. Elizabeth has returned to western Guatemala, where she's bringing computers to school-children. Brian is starting college in western New York. I'm settling in to a new life on the West Coast, thousands of miles from either of them. For the first time since they were born, I have no idea when or where I will see my children next.

I will get used to this phase: the separation, the uncertainty. For now, I pore over my vacation photos. I'd love to show them to you. Of course you'll never see Brian and Elizabeth as I do. But oh, what splendid human beings! I mean, look at them. Would you just look at them?

With this ending, I allowed myself both a "but oh" and an exclamation point. Normally, I let words pull their own weight rather than prop them up with exclamatory punctuation, interjections, italics, or any other external cues. But here, emotion had to spill over. Sometimes the context calls for pulling out the stops. If you do it rarely, if you reserve the special effects for special occasions, you have reason to hope for a powerful impact.

Summary

When you write powerfully, you make countless decisions about words, sentences, paragraphs, and the overall piece. You also make seemingly countless *types* of decisions. Here's a summary of the types that this chapter touches on—a mere sampling of the full range of possible decisions.

- **Topic:** When you have a choice, write about something you know and love.

- **Audience and purpose:** Have an audience and a purpose in mind. Sounds simple, but too many projects fail because of flawed answers to the questions, for whom? and why?

- **Genre:** Pick a genre that best suits your audience and purpose. Master that genre's conventions, if only to equip yourself to reject them intelligently.

- **Organization scheme:** Consider various possibilities before you decide.

- **Tone and diction level:** Make appropriate choices. Stick with them.

- **Title:** Write your title after your theme has emerged. (If you can't state your theme, it hasn't emerged.)

- **Opening strategy:** Jump in to the middle of a story, or find some other way to grab your reader at word one.

- **Word choice:** Choose the best word. (This process involves asking at least five questions and doing the research required to answer them.) Congratulate yourself. Repeat.

- **Imagery:** Use concrete language that appeals to all the senses.

- **Exposition:** Sneak in any necessary background information.

- **Speech:** Use quotations or dialogue to keep readers' attention and to convey character. No writing packs more oomph than the right words wrapped in quotation marks.

- **Transitions:** Create smooth, natural transitions by repeating key words. Place the repeated words close to their predecessors.

- **Trade-offs:** Accept that sometimes one writing guideline prevents you from following another. Choose your trade-offs.

- **Sentence length:** Vary it.

- **Sentence structure:** Vary it.

- **Parallelism:** Use similar structures when you want to emphasize similarities.

- **Action:** Keep things happening.

- **Emotion:** Readers feel most keenly the emotions they fill in for themselves based on what Hemingway calls the "sequence of motion and fact." Use external cues (italics, exclamation points, interjections) rarely; let the words do the work.

- **Motifs:** Use recurring elements to reinforce a theme.

- **Verbs:** Use verbs that punch. Stick with present tense for immediacy, moving backward or forward through time smoothly as needed. Use active voice everywhere except where passive contributes more value.

- **Terms that require definition:** Use obscure terms (from English or any other language) sparingly. Define them as unobtrusively as possible.

- **Punctuation:** Follow the rules, understanding that context sometimes plays a role even in punctuation decisions.

- **Conciseness:** Look hard; you'll find more to cut.

- **Humor:** Let humor happen, unforced, if you see opportunities.

- **Cadence:** Use sound and rhythm—especially in openings and closings—to enhance meaning and to bring attentive readers some measure of pleasure.

- **Endings:** Put key words at the ends of sentences, key sentences at the ends of paragraphs, key paragraphs at the ends of sections, key chapters at the ends of books.

Powerful writing. How can I sum it up? If you gave me your business card and asked me to flip it over and jot down the secret to powerful writing—the distilled wisdom contained in this book—so you could tack it up on your wall, I would write this: "Ask each word: *Why* are you here? Why are *you* here? Why are you *here*?"

Afterword

Discovery consists of seeing what everybody has seen and thinking what nobody has thought.
—ALBERT SZENT-GYORGYI, *THE SCIENTIST SPECULATES*

If you have read this whole book, perhaps you wonder why I don't talk about *that* vs. *which*, or subject-verb agreement—how can I say nothing about subject-verb agreement?—or the clever use of puns, or irony, or narrative point of view (third-person omniscient, etc.), or the difference between predicates and complements, or the en dash, or the author-narrator gap, or the debate over serial commas [*insert serial comma here*] or any number of other topics. I didn't set out to create a comprehensive reference. I think of this book as something like a science center. A guide in a science center might say, "You've got to see how a piezoelectric sensor works! Check out the echo chamber!" I'd say, "Don't these metaphors make your heart sing? Right this way for some well-formed definitions! Please, pick up what you see. Try these things out. You'll like them."

If, on the way out, you were to say that you wish this center had an animal-habitat flannel board or a giant magnet, I'd nod my head. Yes, people need to know about habitats, magnets, and lots of other things. How to justify leaving so much out? How to explain what I've chosen to include?

I wanted you to have a good time while discovering (or rediscovering) a sampling of information you'd find worthwhile. To this end, I've chosen topics that speak to me. Topics that reveal, either explicitly or by example, my own biases and discoveries. Topics that add to what others have said. Topics that, taken together, hint at the range of issues that powerful writers must consider—from the tactical to the strategic, from the practical to the theoretical, from the smallest marks of punctuation to the largest questions of audience and purpose.

> **You have language to command: sentences and paragraphs to craft, and splendid things to build from them.**

I hope that you enjoyed yourself here. I hope that you walk away feeling encouraged and inspired to write powerfully. You have metaphors and definitions to create. You have thoughts to express, visions to convey, processes to document, jokes to tell, dissertations to dissert, proposals to pitch, speeches to deliver, scenes to script, concepts to clarify. You have language to command: sentences and paragraphs to craft, and splendid things to build from them. You have people to connect with—people who need to hear the words that only you can say.

Appendix: Up with Human-Crafted Indexes

A quality index is the user interface to the book.
—Lori Lathrop, Indexing Skills workshop

Indexing is writing.

The indexer doesn't deal in sentences and paragraphs, but, like other writers, the indexer chooses words and arranges them. Like other writers, the indexer reveals structures inherent in a body of information and provides efficient access—ways in—to that information. Like other writers, the indexer accommodates the goals, questions, and vocabularies of a multitude of readers (not to mention the goals, questions, and vocabularies of the same reader on different days). Like other writers, the indexer makes and unmakes many decisions about language. Like other writers, the indexer brings a unique perspective, skill level, and style to each project. Like other writers, the indexer helps to make information understandable and findable. Usable.

In this book, I don't talk about how to index—indexing is a specialized skill with a small group of devotees—but I do want to put in a plug for this kind of writing, which is increasingly undervalued in our Just Google It age. As seasoned technical communicator Mark Baker puts it, "people rarely use indexes for searching anymore, which...means there is less motivation for publishers to pay for good indexing."[176]

Human-crafted indexes need all the plugs they can get.

---

176. Mark Baker comment on "Findability vs. Searchability," *Every Page Is Page One* blog, March 21, 2012, http://everypageispageone.com/2012/03/21/findability-vs-searchability.

If a book or other "closed system"[177] of information has readers who may want to riffle through it for nuggets of knowledge—whether the information inhabits a hardbound book, holes up in an e-reading device, or resides on a honkin' server—those readers deserve a human-crafted index.

My definition of *human-crafted index* includes online indexes, like help-system indexes, or tag-based website maps, or faceted search-and-browse systems. Just as traditional (back-of-the-book) indexers decide which words to use for the entries, which items to list under each entry, and which *see* and *see also* pointers to create, so, too, online indexers decide which words to use for categories and tags, which categories and tags to apply to each page, and which "Related links" pointers to create.

> **Here's what distinguishes a human-crafted index: the technology that generates it takes cues not from bots but from brains.**

Here's what distinguishes a human-crafted index: the technology that generates it takes cues not from bots but from brains.

Imagine, for example, that you want to know what an author has to say about the verb *is*. A typical search engine can tell you every place that *is* appears—perhaps thousands of places—but it can't weed out all the *is*'s that you'd never want to know about. Neither can it guess that you'd also like to know about *are*, *am*, and *be*. A human indexer, on the other hand, would ignore most *is*'s and point only to places that tell you something about *be*-verbs. A human indexer might put those pointers under *is*, *are*, *am*, and *be*.

A human-crafted index represents a knowledge structure, an analysis of topics. When you browse a good index, you learn things that you won't learn any other way. Unlike a table of contents or search engine, a solid index does all of the following:

---

177. For more on open-system vs. closed-system indexing, see Nancy C. Mulvany, *Indexing Books*, 2nd ed. (Chicago: University of Chicago Press, 2005), 4–5.

- Traces the thread of every lookup-worthy topic, large and small, through the entire body of information, skipping over waste-of-time passing references. (The best guideline I've heard for determining lookup-worthiness, or indexability, comes from Lori Lathrop, a past president of the American Society for Indexing. As Lathrop explains in her workshops, when she's indexing, she asks herself with each new entry, would some conceivable reader who jumps from this entry to this content be "happy to be here"? If so, she creates the entry. If not, she doesn't.)

- Tells you at a glance which topics are touched on—and which are whaled on.

- Uses clarifying phrasing (not too much, not too little) to provide context for key words.

- Points you to related topics—without sending you in empty circles.

- Tells you how the key terms in your mind line up with synonyms or related terms in the content.

- Orients you before you read, helps you navigate as you read, and gives you perspective after you've finished reading.

- Points you to topics you want—and to topics you don't yet know you want.

- Helps you decide whether the content is worth your cash.

If you peeked at this book's indexes before you bought the book, you know what I'm talking about. If you haven't yet seen this book's indexes—or bought the book—by all means flick, flip, tap, or scroll your way to them. Why else do you have an index finger?

Resources for Indexers

American Society for Indexing (http://www.asindexing.org)

Beyond Book Indexing: How to Get Started in Web Indexing, Embedded Indexing, and Other Computer-Based Media edited by Diane Brenner and Marilyn Rowland

The Chicago Manual of Style by University of Chicago Press (See the "Indexes" chapter.)

"Editing an Index" by William L. Collins & Karen J. Hamilton, *Intercom: The Magazine of the Society for Technical Communication*, Feb. 2001

Facing the Text: Content and Structure in Book Indexing by Do Mi Stauber

Indexing: A Nuts-and-Bolts Guide for Technical Writers by Kurt Ament

Indexing Books by Nancy C. Mulvany

Inside Indexing: The Decision-Making Process by Sherry L. Smith and Kari Kells

Managing Your Documentation Projects by JoAnn T. Hackos (See the "Managing Indexing" chapter.)

Microsoft Manual of Style for Technical Publications by Microsoft Corp. (See the "Indexing and Attributing" chapter.)

Read Me First! A Style Guide for the Computer Industry by Sun Microsystems, Inc. (See the "Indexing" chapter.)

Science & Technical Writing: A Manual of Style edited by Philip Rubens (See the "Creating Indexes" chapter.)

Search Patterns by Peter Morville and Jeffery Callender

"Single-Source Indexing" by Jan C. Wright (an article on creating a single set of index markers that work for both print and online output, http://www.wrightinformation.com/wrightindex.pdf)

Under the Cover

An early concept for this book's cover.

Early drafts of this book went by the title *The Pen Is Mightier Than the Shovel*, borrowed from the chapter of the same name. For a while, I envisioned some version of this sketch gracing the cover. I include this sketch here because, like the other illustrations that Brian Hull created for this book, it makes me smile. I include it also because, aside from its content, its mere existence reinforces some of my themes: the evolutionary nature of writing, the connection between writing and sketching, and the value of collaboration. Most of all, I include this sketch because it captures the spirit of empowerment that moved me to write this book in the first place. I hope that this spirit stays with you long after you close these covers.

About the Author

Photo by Stella Robertson
Marcia first noting the need for a new book on writing.

When Marcia was twelve, *American Girl* magazine printed her eight-paragraph story, "The Key," and paid her $15. She has been writing ever since. At Lake Forest College, she wrote one-act plays that were performed on the campus stage, learned from, and buried. She studied under Raymond Carver and Tobias Wolff in the Syracuse University creative-writing program. She taught technical writing in the Engineering School at Cornell University. She has done writing of all kinds for organizations of all kinds, from the Fortune 500 to the just plain fortunate.

Marcia has written for the scholarly journal *Shakespeare Quarterly*, the professional journal *Technical Communication*, and the weekly newspaper *Syracuse New Times*. She used to write letters by the boxful. She has contributed posts to her daughter's Peace Corps blog, texts to her son's Droid, and answers to her husband's crossword puzzles. Her words have landed on billboards, blackboards, birthday cakes, boxes of eggs, and the back of this book. She lives in Portland, Oregon.

Glossary

Any given word is a bundle, and meaning sticks out of it in various directions.

—Osip Yemilyevich Mandelstam, *Conversation about Dante*

Bookmark this glossary: www.howtowriteeverything.com/glossary

active voice. The voice in which the subject performs the verb's action instead of receiving it. In *Edgar shoveled that crooked sidewalk four times before noon*, the verb *shoveled* is in the active voice because the subject (*Edgar*) performs the action. If shoveling isn't active, I don't know what is. Compare **passive voice**.

adjectival. Any word or phrase that acts as an adjective. In *Call me a shoveling fool from Liverpool*, the word *shoveling* is an adjectival because, although it's a verb in form (it ends in *-ing*), it acts as an adjective, modifying the noun *fool*.

adjectival compound. See **compound modifier**.

adjective. To call any word an adjective is ambiguous. Is it an adjective in form? In function? Both?

- **Adjective in form:** A form-class word (*crooked*) that can change form, in natural usage, in ways characteristic of adjectives. In other words, an adjective in form is a word with adjective features of form. In isolation, it can pass linguistic tests for adjectiveness. *Crooked*, the standalone word, qualifies as an adjective in form (example test: *crooked+est* = superlative). Of course, *crooked* also qualifies as a verb in form (*crook+ed* = past tense); like many English words, it belongs to multiple form classes.

- **Adjective in function (an adjectival):** Any word or phrase that acts as an adjective (a modifier of a noun) in a phrase or clause. An adjective in function typically describes something. In *Siegrid cleared the crooked sidewalk*, the word *crooked* is an adjective not only in form but also in function because it modifies *sidewalk*.

> To call any word an adjective is ambiguous. Is it an adjective in form? In function? Both?

advance organizer. A preview or overview. Advance organizers typically describe the structure of the information to come, sometimes listing the section headings as a sectional table of contents. This device presumably gets its name from its purpose: organizing the reader's brain in advance of reading.

adverb. To call any word an adverb is ambiguous. Is it an adverb in form? In function? Both?

- **Adverb in form:** A form-class word (*frantically*) that has adverb features of form. In isolation, an adverb in form can pass linguistic tests for adverbness. *Frantically*, for example, has the telltale *-ly* ending, so you might call it an adverb in form. You can't be sure with adverbs, though. They are "the most difficult of the four form classes to identify" by form alone because adverbs and adjectives have overlapping form characteristics.[178] (The adjective *friendly* ends in *-ly* too.)
- **Adverb in function (an adverbial):** Any word or phrase that acts as an adverb (a modifier of a verb, an adjective, or another adverb) in a phrase or clause. An adverb in function typically tells when, where, or how something happens. In *Tim shoveled frantically*, the word *frantically* is an adverb not only in form but also in function because it describes the manner in which Tim did his shoveling.

adverbial. Any word or phrase that acts as an adverb. In *Zelda hurled her shovel into the ravine*, the prepositional phrase *into the ravine* is an adverbial because it tells us the direction in which Zelda did her hurling.

alliteration. The repetition of sounds within words or among neighboring words. Alliteration comes in two types: assonance and consonance. In *Utterly, unutterably sumptuous to utter*, the alliteration consists of six *uh*s and five *t* sounds. (Only the sounds count; the ear doesn't care about spelling.) The *New York Times* crossword puzzle often mixes both types of alliteration in a clue, as in *Stash for cash* (answer: "IRA").

Used judiciously, alliteration adds pizzazz. Too much alliteration distracts the reader and sounds corny.

---

178. Klammer, Schultz, and Della Volpe, *Analyzing English Grammar*, 81.

amplification. The repetition of a word or phrase followed by additional detail. Example: *The sidewalk, groaning with snow, aroused Gena's sense of responsibility—her sense of obligation, decency, and saintliness.* The repetition of the phrase *sense of* affords the addition of three nouns—*obligation*, *decency*, and *saintliness*—that amplify the meaning of *responsibility*.

Here's another amplification example (from *"Explore and Heighten:* Magic Words from a Playwright" on page 103): *Those are the times to add detail, the times to expand.* The repetition of *the times* affords the addition of a second phrase (*to expand*) that amplifies the first (*to add detail*).

anadiplosis. The repetition of words from the end of one sentence or clause at the beginning of the next: *Clint decided that the time had come. The time had come to haul out the snow blower.* This device can help create emphasis or transition.

Here's another anadiplosis example (from "Decisions, Decisions" on page 161): *This usage resonates with me—its upside-downness. Upside-down is how I feel in this place.*

analogy. A comparison that highlights similarity not just between two things but also between two relationships: *That kid is as handy as a pocket on a shirt.* (Relationship 1: Pocket on a shirt and handiness. Relationship 2: Kid and—by extension—handiness.) Analogy goes beyond straight metaphor, which would have the kid *being* a pocket on a shirt.

At its most useful, the by-extension part of an analogy illuminates the unfamiliar. At its least useful, it creates a logical fallacy, implying that the by-extension similarity equals truth: *Just as efforts to influence the weather are futile, so, too, are efforts to influence language usage.* Analogous reasoning can have a persuasive effect on people who fail to detect the points at which the analogy breaks down. For a discussion of this weather-language analogy, see "Up with (Thoughtful) Prescriptivism" on page 7.

anaphora. The repetition of a key word or phrase at the beginning of consecutive clauses or sentences: *The snow fell for an hour; the snow fell for a day; the snow fell for weeks—and it's still falling.*

For another example of anaphora, see the opening paragraph of "Appendix: Up with Human-Crafted Indexes" on page 183. Five consecutive sentences start with *Like other writers*.

anecdote. A short, usually true story that introduces, clarifies, or reinforces what's being said. Want to draw your reader in? Start with *The other day.* For an example, see the opening of "Who's Your Sam?" on page 153.

antecedent. The noun or noun phrase to which a pronoun refers. In *Where's my parka? I know it's around here somewhere*, the noun *parka* is the antecedent for the pronoun *it*. Keeping pronouns close to their antecedents avoids ambiguity.

anthimeria (antimeria). See **enallage**.

antithesis. The juxtaposition of two contrasting ideas highlighted by a grammatically parallel structure: *To ignore the snow is human; to shovel, divine.*

Here's another antithesis example (from "Decisions, Decisions" on page 161): *She persisted. I relented.*

appositive. A noun or noun phrase that renames the noun directly preceding it. In *My neighbor Aleks, a rock collector, is digging out her driveway,* the phrase *a rock collector* is an appositive. If an appositive is nonrestrictive (not required to identify the noun), it is set off with enclosing commas.

article. See **determiner**.

aspect. A verb attribute similar to tense in that it conveys information about time. In English, aspect and tense are tangled up together. Aspect has to do with an action's ongoingness or lack thereof: *Bob is/was shoveling* or *Bob has/had shoveled.* You need terms like *progressive* and *perfect* to talk about aspect. I leave it to you, if you're so inclined, to venture into those deeps. Compare **mood, tense, voice**. See also **auxiliary**.

assonance. A type of alliteration in which vowel sounds are repeated, as in *Come up to Uncle Bud's for supper.* Here, the assonance consists of five *uh*s. (Only the sounds count; the ear doesn't care about spelling.) The *New York Times* crossword puzzle often uses assonant clues, as in *Prepare to share* (answer: "divide").

auxiliary. A structure-class word that signals the coming of a main verb. In *Let's get working,* the auxiliary *get* signals the coming of *working.* In *Teresa will have gone ice skating by this time tomorrow,* the auxiliaries *will* and *have* signal the coming of the main verb *gone.*

Although sometimes called *auxiliary verbs* or *helping verbs,* auxiliaries are not "true verbs."[179] They are verb helpers. Unlike true (main) verbs, auxiliaries are not form-class words. We recognize auxiliaries not by form but by function. One or more auxiliaries work with the main verb to determine tense, voice, mood, and aspect.

The role of auxiliary can be played by a *be*-verb, *have*-verb, *do*-verb, or *get*-verb or by one of the modal auxiliaries: *can, could, will, would, shall, should, ought, may, might, must.*

Unlike the modal auxiliaries, *be*-verbs, *have*-verbs, *do*-verbs, and *get*-verbs can also function as main verbs: *I'll be fine. You'll get better.* In the leading role, *be, have, do,* and *get* transform from verb-helping auxiliaries (structure-class words) into true verbs (form-class words).

---

179. Klammer, Schultz, and Della Volpe, "Auxiliaries" in *Analyzing English Grammar,* 106–111.

Unique among the auxiliaries, the modal auxiliaries, in normal usage, play a supporting role every time. You never hear of people *mighting* their hearts out, for example, although they might sing their hearts out. Linguists argue convincingly that modal auxiliaries never act as form-class words—as main verbs—even when they seem to. Modal auxiliaries have no features of form; they never change form. *Might* is *might* is *might*. For these reasons, modal auxiliaries qualify as "prototypical structure words";[180] in normal usage, they never stray from the structure class into the form class. Unless they take a notion to act as nouns. They do have the might. Compare **linking verb**.

bdelygmia. A series of exuberantly disparaging remarks: insult as art form. *Bdelygmia* (de-LIG-me-uh) comes from the Greek word for "abuse." Fans of the movie *Monty Python and the Holy Grail* will recognize this example spewn by a Frenchman from a battlement to the foreigners below:

> You don't frighten us, English pig-dogs! Go and boil your bottoms, sons of a silly person! I blow my nose at you, so-called Ah-thoor Keeng, you and all your silly English K-n-n-n-n-n-n-niggits! ... You empty-headed animal food trough wiper! I fart in your general direction! Your mother was a hamster and your father smelt of elderberries![181]

be-**verb.** Any form of the verb *to be*—*am, are, be, been, being, is, was, were*—whether it acts as an auxiliary (*You are getting tired*) or as a main verb (*You are tired*). As an auxiliary, a *be*-verb is a structure-class word, not a "true verb."[182] As a main verb, a *be*-verb is a form-class word: a true verb. See also **linking verb**.

bossy verb. See **imperative**.

brainstorm.

- A natural phenomenon that occurs in a room when a group of people fill the air with hundreds of charged ideas.
- A natural phenomenon that occurs in individuals' heads as they write, sketch, or create anything, causing them to forget to eat, sleep, and change out of their slippers when they go outside.

clause. A group of related words containing a subject and a verb. Clauses come in two types: independent and dependent (subordinate). Compare **phrase**.

---

180. Ibid., 107.
181. "Monty Python and the Holy Grail," *Wikiquote*, last modified March 6, 2012, http://en.wikiquote.org/wiki/Monty_Python_and_the_Holy_Grail.
182. Klammer, Schultz, and Della Volpe, *Analyzing English Grammar*, 108–109.

comma-spliced sentence. A run-on sentence that includes two independent clauses joined by only a comma, making the comma a comma splice: *Max bent down to pick up the rock, he heard his back snap.* Compare **fused sentence.**

complex sentence. A sentence that contains exactly one independent clause and at least one dependent clause: *Although the snow blower's lack of cooperation frustrated him, Xavier persisted until he got the thing running again.* A complex sentence distinguishes the subordinate ideas (Xavier's frustration) from the more important ones (Xavier's persistence).

compound-complex sentence (complex-compound sentence). A sentence that contains two or more independent clauses and at least one dependent clause: *Although the snow blower's lack of cooperation frustrated him, Xavier persisted until he got the thing running again, and the neighbors cheered.*

compound modifier (adjectival compound, phrasal adjective, unit modifier). A phrase that functions as a unit in modifying a noun. When a compound modifier precedes the noun, it requires a hyphen—with rare (and hotly disputed) exceptions. In *snow-crusted chronicles*, the words *snow* and *crusted* form a compound modifier, acting as an adjective that describes, oddly if mellifluously, the noun *chronicles*.

compound sentence. A sentence that contains two or more independent clauses: *Xavier's snow blower conked out, and he had a devil of a time getting it to run again.*

conjunction. A structure-class word that joins words, phrases, or clauses. Conjunctions come in various types, including coordinating conjunctions, subordinating conjunctions, and conjunctive adverbs.

conjunctive adverb. An adverb that acts as a conjunction: *therefore, nevertheless, however, subsequently, otherwise, then.* Conjunctive adverbs join two independent clauses, signaling a relationship between them: cause and effect, sequence, contrast, comparison, etc.

In *Siobhan shoveled as fast as she could; nevertheless, night fell before she could finish the job*, the word *nevertheless* acts as a conjunctive adverb, revealing a contrast between Siobhan's speed and the sun's. A conjunctive adverb is followed by a comma and preceded by a semicolon—or a period: *Siobhan shoveled as fast as she could. Nevertheless, night fell.*

(The words identified here as conjunctive adverbs may also play other roles, in which case they are classified differently.)

Compare **coordinating conjunction** and **subordinating conjunction.**

consonance. A type of alliteration in which consonant sounds are repeated, as in *Tuckered-out and thirsty, Titus can't feel his toes.* Here the consonance consists of seven *t* sounds. (Only the sounds count; the ear doesn't care about spelling.) The *New York Times* crossword puzzle often includes consonant clues, like *Parmesan pronoun* (answer: "mio," Italian for *my*).

> The *New York Times* crossword puzzle often includes consonant clues, like *Parmesan pronoun* (answer: "mio," Italian for *my*).

content words. See **form-class words**.

coordinating conjunction (coordinator). A conjunction that joins words, phrases, or clauses of equal grammatical rank or function (coordinate words, phrases, or clauses). Coordinating conjunctions have no characteristic features of form. You can remember the most common ones by the acronym FANBOYS: *for, and, nor, but, or, yet, so.*

In *Hey, fanboy, how about a bowl of chili or a mug of wassail?* the coordinating conjunction is *or. A bowl of chili* and *a mug of wassail* are coordinate phrases.

In *Jacqueline drove through the storm to get chili powder, and Jacques stayed home to cook,* the coordinating conjunction is *and,* and the two independent clauses are the coordinate elements. When the coordinate elements are independent clauses, the conjunction is preceded by a comma.

As for the notion that sentences shouldn't start with *and* or *but,* forget it—lest you fall for a "rank superstition" and a "gross canard."[183] Kicking off with *and* or *but* is, in fact, "highly desirable in any number of contexts." Good writers do it all the time, especially in informal writing. But—make that *and*—they don't follow either little word with a comma.

(The words identified here as coordinating conjunctions may also play other roles, in which case they are classified differently.)

Compare **conjunctive adverb** and **subordinating conjunction**.

copular verb. See **linking verb**.

dangling modifier. A word or phrase intended to modify a word that's missing …gone… *verschwunden.* When you calibrate your eye to danglers, a new source of humor opens up for you. In *As a mother of eight, my sidewalk is*

---

183. Garner, *Garner's Modern American Usage,* 44 (under "and"), 121 (under "but").

never shoveled, the word *mother* is the dangling modifier. It's intended to modify *I,* but the sentence contains no such word. *Mother* is left dangling. It has no choice but to modify the only noun in sight, *sidewalk,* creating a ridiculous pairing—unless the sidewalk has, in fact, spawned eight little sidewalks. Compare **misplaced modifier** and **squinting modifier**.

dependent clause (subordinate clause). A clause that depends on an independent clause to form a complete sentence. In *People skittered off to the sides of the road because a neophyte driver was barreling down the icy street,* the second half of the sentence—*because a neophyte driver was barreling down the icy street*—is a dependent clause.

> **Mother is left dangling. It has no choice but to modify the only noun in sight, *sidewalk,* creating a ridiculous pairing—unless the sidewalk has, in fact, spawned eight little sidewalks.**

A dependent clause begins with a subordinating conjunction (in this case, *because*). By itself, a dependent clause is a sentence fragment. A dependent clause requires no punctuation when it follows the independent clause. When the dependent clause comes first, it is followed by a comma: *Because a neophyte driver was barreling down the icy street, people skittered off to the sides of the road.*

Compare **independent clause**.

determiner. A structure-class word—*a, an, the, this, that, those, my, her, his, its, their, every, many, one, two, second, last* (an article, a possessive, a number, etc.)—that precedes and modifies a noun but is neither an adjective nor another noun. Examples: *this task, their travails, every livelong day.* (The words identified here as determiners may also play other roles, in which case they are classified differently.)

deverbal -*ing* noun. An -*ing* noun with no verb qualities beyond the superficial resemblance. In *Every cloud has a silver lining,* the word *lining* has only noun qualities: it's a direct object, and you could replace it only with another noun (like *layer*); you could not replace it with an infinitive (*to line*). Compare **gerund**.

diction level. The degree of formality in word choice. *I beg your pardon, excuse me,* and *say what?* say the same thing at various levels of diction.

direct object. A noun that completes the meaning of a transitive verb, answering the question, what? In *Tony whacked the snowbank,* the noun *snowbank* is the direct object of the transitive verb *whacked.*

dummy word. See **expletive**.

enallage (anthimeria, grammatical shift). Usage of a word outside its natural forms or functions. In *This weather will not peace us,* the word *peace* functions, uncharacteristically—enallagistically, you might say— as a verb.

E-Prime (English-Prime, E´). A form of English that excludes *be*-verbs. Advocates claim that E-Prime (proposed by D. David Bourland Jr., a student of philosopher Alfred Korzybski) clarifies thinking and strengthens writing. E-Prime rejects statements like *Shoveling is the worst,* which presents judgment as fact, in favor of statements that more accurately communicate a speak-er's experience: *I spit in shoveling's general direction.*

> E-Prime rejects statements like *Shoveling is the worst,* which presents judgment as fact, in favor of statements that more accurately communicate a speaker's experience: *I spit in shoveling's general direction.*

equational verb. See **linking verb**.

exclamation. See **interjection**.

expletive (dummy word). A word that has no grammatical function. In phrases like *there is, there are, it is,* and *it was,* the words *there* and *it* are expletives. According to the *Oxford English Dictionary, expletive* (in its adjective form) means "introduced merely to occupy space…serving merely to fill out a sentence, help out a metrical line, etc. Also occas. of a mode of expression: Redundant, wordy." See also **filler word**.

figure of speech. A colorful expression with idiomatic meaning, a turn of phrase: *Dog my cats!* ("I'll be dipped!"). Compare **rhetorical device**.

filler word. A word that contributes no meaning—and, therefore, typically no value—to a phrase or sentence.

Two types of filler words to delete (usually) are qualifiers and expletives. In *Vera feels somewhat cold,* the qualifier *somewhat* adds no value. In *There is no reason to turn the heat down,* the expletive *there* adds no value. Better: *Vera feels cold. Don't turn the heat down.*

A filler word may add value in terms of meter or sound. The Nat King Cole lyric "V is very, very extraordinary" would be tough to sing without the filler *verys*. And we'd have nothing to smile about without the *some-what* here: Shirts that have haikus / Are somewhat overrated / Still I'm wearing one.

foot. See **metrical foot**.

form. A word's physical shape, the aspect of the word that you see or hear. Every word has form. The form of the word *sidewalk* is s-i-d-e-w-a-l-k. If you add or delete or change the letters—whether meaningfully (*sidewalks*) or randomly (*sidewalkqwerty* or *qwertysidewalk* or *sdqwertywks*)—you change its form. *Going* (*go+ing*) is a form of the verb *to go*. *Went* is also a form of *to go*. Because *went* falls outside the standard pattern of verb conjugation (past tense = <verb>+ed), *went* is called an irregular form of the verb. Compare **function**.

form-class words (content words, parts of speech). Words in any of the form classes: nouns (*house*), verbs (*welcomed*), adjectives (*warm*), and adverbs (*warmly*). Modern linguists consider these classes—only these four—the parts of speech. Form-class words, or content words, usually contain not grammatical meaning (as structure-class words do) but lexical meaning, that is, meaning in themselves.

Form-class words, the majority of English words, have something in common that sets them apart from words in the structure classes: they generally change form in characteristic ways. In isolation (out of context), these words can be linguistically tested in ways that help classify them. For example, adding an *s* transforms a noun into a plural that English speakers would use naturally (*house+s* = *houses*), and adding *est* transforms an adjective into a superlative (*warm+est* = *warmest*).

When a word changes form in ways characteristic of a given form class, linguists call that word a noun, verb, adjective, or adverb in form.

While some words never stray from a single form class—*desk*, for example, is a prototypical noun (you wouldn't normally say, "We're desking" or "That's the deskest")—many English words can belong to multiple form classes. *House* belongs to two form classes: nouns and verbs. It changes form in ways characteristic of nouns (*houses, house's*), so it qualifies as a noun in form—and it changes form in ways characteristic of verbs (*housed, housing*), so it also qualifies as a verb in form.

Self-proclaimed enigmatologist Will Shortz, crossword-puzzle editor for the *New York Times* and consummate creator of duplicitous clues, has built his following on the backs of words like this—words that move easily between form classes. Take the tease of a clue *Defeat in a derby*. Are we meant to read *defeat* as a noun (as in "a defeat in a derby") or as a verb ("to defeat in a derby")? The clue alone lacks sufficient context. We have to fill in some neighboring answers to determine the answer: "outride" (a verb—aha!). In some puzzles, the same clue appears multiple times, yielding answers from multiple form classes. The clue *mean* might appear twice, yielding a noun ("average") for one answer and a verb ("signify") for another. In crossword puzzles, as in everyday usage, a form-class word holds clues within itself but reveals its full meaning only in the context of other words.

fragment. See **sentence fragment**.

function. The grammatical role that a word or phrase plays in a phrase or clause. *Sidewalk* functions as (acts as) a direct object in *Let's shovel this sidewalk* and as an adjective in *I've got the sidewalk blues.* Just as a musical note's function in a chord is determined by its position relative to the other notes—the same note contributes to a major chord here, a minor chord there—a word's or phrase's function in a phrase or clause is determined largely by its position relative to the other words. Compare **form**.

function words. See **structure-class words**.

fused sentence. A run-on sentence that includes two independent clauses joined by only a space: *Max bent down to pick up the rock he heard his back snap.* Compare **comma-spliced sentence**.

gerund. An *-ing* noun with verb qualities (a verbal noun). Some linguists define the gerund the other way around: a verb that acts as a noun (a nominal verb). Either way, in *Lining clouds with silver is no easy job,* the word *lining* is a gerund because it has both noun qualities (it's the subject of the sentence) and verb qualities (you could replace it with the infinitive *to line*—a sure-fire test of gerundness). Compare **deverbal -ing noun**.

grammatical shift. See **enallage**.

helping verb. See **auxiliary**.

hyperbole. The use of exaggeration to create emphasis or effect: *An Alpine avalanche hit our house last night.*

iamb (iambus). A type of metrical foot. An iamb is an unaccented syllable followed by an accented syllable: da-DUM. *Prefer* is an iamb. *Come on!* is an iamb. A series of iambs creates an iambic pattern.

idiom. A word or combination of words whose commonly understood meaning differs from the literal meaning. In *Keith did a bang-up job chipping the ice off the living-room windows,* you can rest assured that no banging was involved. See also **phrasal verb**.

imagery. Concrete language, language that appeals to the senses—all of them. (Of all words, how has *imagery*—*image* with a tail—come to mean not just what we see but also what we smell, hear, touch, and taste?) Psychologists claim that we process concrete language more quickly than abstract language—and, further, that the faster we process the words, and the more our senses tingle along the way, the more likely we are to believe what we read.[184] No word from psychologists on the increased likelihood of enjoying it.

imperative (bossy verb). A verb's command form. In *Help me knock these icicles off the gutters*, the verb *help* is an imperative. Technically, imperative is one of several verb moods.

independent clause. A clause that stands alone—or could—as a complete sentence. In *People skittered off to the sides of the road because a neophyte driver was barreling down the icy street*, the first half of the sentence— *People skittered off to the sides of the road*—is an independent clause. Compare **dependent clause**.

infinitive (*to*-verb). The *to* form of a verb: *to chop, to scrape, to fling, and not to yield*. (Apologies to Alfred Lord Tennyson.)

***-ing* noun.** A word ending in *-ing* that functions as a noun: either a gerund or a deverbal noun.

intensifier. A qualifier that supposedly intensifies another word but doesn't. Examples: *very, such, so*.

interjection (exclamation). A word or phrase, often placed at the beginning of a sentence, used to express emotion or to indicate voice: *ah! hi, oh, well, um, hey, wow! that's great!* Traditionally, the interjection has been considered a part of speech, but it qualifies as neither a form-class word nor a structure-class word. It's a grammatical outlier, like the expletive. Use interjections rarely, but don't rule them out. Sometimes you need a good *yikes!*

interrogative. A structure-class word used to begin a question: *who, whom, whose, which, what, where, when, how, why*, etc.

(The words identified here as interrogatives may also play other roles, in which case they are classified differently.)

intransitive verb. See **transitive and intransitive verbs**.

linguistics. The scientific study of human language.

linking verb (equational verb, copular verb). A *be*-verb or other verb—*seem, appear, become, remain, grow, get*—that equates two elements in a clause.

---

184. Jochim Hansen and Michaela Wänke, "Truth from Language and Truth from Fit: The Impact of Linguistic Concreteness and Level of Construal on Subjective Truth," *Personality and Social Psychology Bulletin*, 36, no. 11 (Society for Personality and Social Psychology, November 2010), 1576–1588. Abstract: http://psp.sagepub .com/content/36/11/1576. Discussed in Jeremy Dean, "Why Concrete Language Communicates Truth," *PsyBlog* blog, June 29, 2011, http://www.spring.org.uk/2011/06 /why-concrete-language-communicates-truth.php.

In *The icicles are slippery*, the linking verb *are* equates *icicles* and *slippery*. When a word acts as a linking verb, it is (like the *is* you just read) a form-class word. Compare **auxiliary**.

-ly word. See **adverb**.

metalanguage. Words about words (*noun, preposition, form class, structure class, modifier, independent clause*, etc.). Metalanguage is talk that talks about itself, a downright gymnastic proposition. I would compare it to a sketch sketching itself, but everyone knows that's impossible.

metaphor (comparative trope). A comparison of one thing to another. *This sled has wings* is a metaphorical statement that compares a sled to something that can literally fly. Throw in *like* or *as*, and you have a simile, a type of metaphor whose comparison is explicit: *This sled sails through the air like a* <crash> ... never mind.

meter. The rhythmic structure of a group of words, that is, the patterns formed by their accented syllables. Meter is determined by two elements: the type of metrical foot and the number of feet per grouping. For example, iambs repeated in groups of five form the meter known as iambic pentameter: da-DUM da-DUM da-DUM da-DUM da-DUM.

> **Even if you aren't writing poetry, keep meter in mind as you choose your words— especially at the ends of sentences or sections—to GIVE your READers THAT much MORE to LIKE.**

Even if you aren't writing poetry, keep meter in mind as you choose your words—especially at the ends of sentences or sections—to GIVE your READers THAT much MORE to LIKE.

metrical foot (foot). The basic unit of meter. The best-known metrical foot is the iamb (da-DUM). A metrical foot is the grouping of syllables according to two elements: the number of syllables in the unit and the arrangement of accented syllables. If you were to tap your foot to the beat as you read aloud, you'd tap once per accented syllable but not necessarily once per metrical foot. The longest foot, a dispondee, is DUM-DUM-DUM-DUM. Metrical feet come in over two dozen types, whose names (ditrochee, molossus, etc.) MOST of US will NEVer NEED to KNOW.

misplaced modifier. A word or phrase that, by virtue of its position, modifies the wrong word. In *Geraldine saw the snowplow peeking through the window*, the phrase *peeking through the window* follows, and therefore seems to describe, *snowplow*. Presumably, Geraldine did the peeking, in which case the phrase *peeking through the window* is a misplaced modifier. It belongs next to the word it modifies: *Peeking through the window, Geraldine saw the snowplow.* Compare **dangling modifier** and **squinting modifier.**

modal auxiliary. See **auxiliary.**

modifier. A word or phrase that modifies—adds meaning to—another word. In *wooden handle*, the word *wooden* modifies the word *handle*. For clarity, modifiers must stay as close as possible to the words they modify. Otherwise, the sentence could end up suffering (perhaps hilariously) from one of these modification errors: dangling modifier, misplaced modifier, or squinting modifier. See also **compound modifier.**

mood (modality). A verb attribute that indicates such abstractions as conditionality, probability, obligation, ability. Moods in English include the following:

- indicative (the most common mood: *Taylor goes* or *Taylor is going*)
- imperative (*Taylor, go!*)
- subjunctive (*if only Taylor were going*)
- conditional (*Taylor would go*)

Compare **aspect, tense, voice.** See also **auxiliary.**

motif. An element that recurs throughout a piece of writing. If you read these glossary entries straight through, you'll discover two motifs: winter activities and the *New York Times* crossword puzzle.

nominal. Any word, phrase, or clause that acts as a noun. In *That Suki wanted to take a break didn't stop her from finishing the job,* the clause *That Suki wanted to take a break* is a nominal because it acts as a noun: the subject of the sentence.

noun. To call any word a noun is ambiguous. Is it a noun in form? In function? Both?

- **Noun in form:** A form-class word (*shovel*) that can change form, in natural usage, in ways characteristic of nouns. In other words, a noun in form is a word with noun features of form. In isolation, it can pass linguistic tests for nounness. *Shovel*, the standalone word, qualifies as a noun in form (example tests: *shovel+s* = plural; *shovel+'s* = possessive). Of course, *shovel* also qualifies as a verb in form; like many English words, it belongs to multiple form classes.
- **Noun in function (a nominal):** Any word, phrase, or clause that acts as a noun. A noun in function typically names a person, place, or thing. In *Carl broke the shovel over his knee,* the word *shovel* is a noun not only in form but also in function because it names the thing that Carl broke (grammatically, the direct object).

onomatopoeia.
- The reinforcement of meaning in words' sounds: *buzz, chop, slide, crackle, murmur, mellifluous.*
- A word that has onomatopoetic (also *onomatopoeic*) qualities. Plural: *onomatopoeias.* Say that with a straight face.

paragraph. One sentence or a group of sentences that stands alone as a compositional unit. A paragraph can be defined various ways:
- **By its components:** A paragraph typically contains a topic sentence and multiple supporting sentences.
- **By its content:** A paragraph typically develops one main idea.
- **By its typography:** A paragraph typically begins with an indent, outdent, or simple line break.
- **By its structure:** A paragraph typically develops according to a coherent structure: chronological order, logical progression, spatial sequence, or some other organizational scheme.
- **By its purpose:** A paragraph typically has one of these purposes: to describe, to persuade, to create a desire to turn the page.

To build powerful paragraphs, heed *The Little English Handbook* author, Edward Corbett, who urges writers to "take care" of "the three most persistent and common problems that beset the composition of written paragraphs," namely, "unity, coherence, and adequate development."[185]

parallelism. The repetition of grammatical structure, sound, meter, meaning, etc., within a sentence or from one sentence to another. In *Roland put on his heaviest coat, his thickest gloves, his widest muffler, and his warmest hat,* the structure [*his* <adjective+*est*> <noun>] recurs four times, creating parallelism within the sentence.

particle. See **verb particle**.

parts of speech. Ask a traditional linguist and a modern linguist to name the parts of speech, and you'll get shockingly different lists:
- **Traditional parts of speech:** nouns, verbs, adjectives, adverbs, prepositions, pronouns, conjunctions, and interjections (give or take a part).
- **Modern parts of speech (form-class words):** nouns, verbs, adjectives, and adverbs.

See also **structure-class words**.

passive voice. The voice in which the sentence's subject receives the verb's action instead of performing it. Passive voice is indicated by a passive

---

185. Edward P. J. Corbett, *The Little English Handbook,* 3rd ed. (New York: Wiley, 1980), 85.

marker, namely, the combination of an auxiliary *be*-verb (*was*) and the past-participle (*-ed*) form of the main verb. In *The sidewalk was shoveled*, the subject (*sidewalk*) receives the action. Typical sidewalk if you ask me. Compare **active voice**.

pentameter. A MEter conSISTing of FIVE METrical FEET.

periodic structure. A structure (of a phrase, sentence, paragraph, or section) in which the emphatic information appears at the end. A classical periodic sentence comprises a series of clauses that build to the main clause, leading to a...a...a climax. *No matter how much Mandy begged to stay inside, no matter how loudly she pleaded, no matter how pitifully she wept, her mother—without a single sign of sympathy—continued to insist that she go outside with her friends and play in the snow.*

personal pronoun. A pronoun with attributes related to grammatical person (first-person, second-person, third-person): *I, you, we, she, herself, their, it, its,* etc.

phrasal adjective. See **compound modifier**.

phrasal verb. A multiple-word verb (*chip in, drop out of*) that has an idiomatic meaning, a meaning different from that of the individual words (*chip in* means "help"; *drop out of* means "quit").

The *New York Times* crossword puzzle wouldn't be the *New York Times* crossword puzzle without phrasal verbs. Examples:

- The clue *Give _____ to (approve)* yields the answer "anod." (*Give a nod to* is a phrasal verb meaning "approve.")
- The clue *Long (for)* yields the answer "hope." (*Long for* and *hope for* are synonymous phrasal verbs.)
- The clue *Distribute, with "out"* yields the answer "parcel." (*Parcel out* is a phrasal verb meaning "distribute.")

See also **verb particle**.

phrase. A group of related words that contains no subject-verb relationship: *the neophyte driver* or *at high noon* or *barreling down the icy street*. Compare **clause**.

preposition. A structure-class word (*from, with, over, into,* etc.) that typically appears immediately before—in *pre-position* to—a noun phrase. The preposition connects the noun phrase to another word in the sentence. In *The mug of coffee dissipated welcome warmth into Hubert's frozen fingers*, the preposition *into* connects the noun phrase *Hubert's frozen fingers* back to the verb *dissipated*. The prepositional phrase *into Hubert's frozen fingers* modifies the verb *dissipated*, describing how and where the dissipating happened. The prepositional phrase, as a whole, plays the role of an adverb in this sentence; the word *into* plays the role of a preposition.

(The words identified here as prepositions may also play other roles, in which case they are classified differently.)

pronoun. A structure-class word (*he, she, it, this, those, that,* etc.) that can substitute for a noun or noun phrase. Unique among the structure-class words, pronouns

> **The words identified here as prepositions may also play other roles, in which case they are classified differently.**

can change their form (*he* changes to *him* or *his,* for example). In spite of this commonality with form-class words, though, pronouns are classified as structure-class words because they are defined by their function; they create relationships between words.

Pronouns come in lots of types: personal, relative, reflexive, indefinite, etc.

(The words identified here as pronouns may also play other roles, in which case they are classified differently.)

prototype word. A word that belongs to only one class. *Sidewalk* is a prototype word, a prototypical noun. You would not normally use it as a verb (*sidewalked*), adverb (*sidewalkly*), adjective (*sidewalkest*) or any other type of word.

Shovel is not a prototype word. It works as either a noun or a verb: You need a *shovel* (noun) to *shovel* (verb).

See also **auxiliary, form-class words, structure-class words**.

qualifier. A structure-class word or phrase (*very, such, so, fairly, a bit*) that precedes an adjective or adverb and supposedly either increases or decreases its degree: *very hot, a bit shovel-weary.* (Qualifiers that increase the degree are called intensifiers.) Qualifiers rarely add value. *Hot* and *shovel-weary* work better alone. See also **filler word**.

relative clause. A dependent clause that modifies (and, typically, directly follows) a noun or noun phrase. In *I know the folks whose woods these are* (apologies to Robert Frost), the dependent clause *whose woods these are* is a relative clause that modifies the noun *folks.* A relative clause starts with a relative pronoun.

relative pronoun (a relative). A pronoun (*that, which, who, whose, whom*) that introduces a relative clause. In *I know the folks whose woods these are* (apologies again to Robert Frost), the word *whose* is a relative pronoun. It's called a relative pronoun because it relates the clause (*whose woods these are*) to the noun that the clause modifies (*folks*). Some linguists call these words relatives, not relative pronouns, because in this role they are not substituting for nouns.

(The words identified here as relative pronouns may also play other roles, in which case they are classified differently.)

See also **remote relative**.

remote relative. A relative pronoun that is positioned too far away from its antecedent. In *Gunther finally dug up his potatoes whose field almost froze before he got around to the task*, the word *whose* is a remote relative—too far away from its antecedent, *Gunther*. Because *whose* follows *potatoes*, the field seems at first to belong to the spuds. Here's one fix: *Gunther, whose field almost froze before he got around to the task, finally dug up his potatoes.*

restatement. A rhetorical device in which an idea is repeated in a series of synonymous phrases or statements. Here's an example from "*Explore and Heighten:* Magic Words from a Playwright" on page 103: *Those are the times ... to expand. Build up. Pile on the voom.* Restatement, like many rhetorical devices, creates emphasis. Pile on the restatement!

rhetorical device. A technique that a writer or speaker uses (alliteration, metaphor, hyperbole, etc.) to clarify, emphasize, persuade, delight, or otherwise engage the reader.

rhetorical question. A question that either can't be or isn't intended to be answered. A rhetorical question emphasizes a point while letting the writer's voice come through. How many rhetorical questions do you suppose this book contains?

run-on sentence (run-on). A sentence that includes two independent clauses joined insufficiently. Run-on sentences come in two types: comma-spliced sentences (*Max bent down, he heard his back snap*) and fused sentences (*Max bent down he heard his back snap*).

Some run-ons serve a purpose, as when several short independent clauses form a unit (*Max came, he saw, he ran for cover*). Most of the time, though, a run-on sentence of either type reads better when punctuated with one of the following:

- A period, semicolon, colon, or dash: *Max bent down; he heard his back snap.*
- A coordinating conjunction preceded by a comma: *Max bent down, and he heard his back snap.*
- A conjunctive adverb preceded by a semicolon and followed by a comma: *Max bent down; simultaneously, he heard his back snap.*

sentence. One word or a group of words that stands alone as a grammatical unit. A sentence can be defined in various ways:

- **By its components:** A sentence typically contains at least one subject and a related verb.
- **By its content:** A sentence typically forms a complete thought.
- **By its typography:** A sentence typically begins with a capital letter.
- **By its punctuation:** A sentence typically ends with a period, question mark, or exclamation point.
- **By its structure:** A sentence typically has one of these structures: simple, compound, complex, or compound-complex.
- **By its purpose:** A sentence typically has one of these purposes: to state (declarative sentence), to command (imperative sentence), to ask (interrogative sentence), to exclaim (exclamatory sentence).

No definition of *sentence* captures all the flexibility and creative potential of this fundamental unit of grammar.

Compare **clause, phrase, sentence fragment**.

> **No definition of *sentence* captures all the flexibility and creative potential of this fundamental unit of grammar.**

sentence fragment. A phrase or clause that is punctuated as if it were a sentence but that does not stand alone grammatically as an independent clause. Like this.

In informal writing, when used judiciously, sentence fragments can enhance the reading experience by creating emphasis, suspense, and variety. Keep fragments short so readers won't mistake them for complete sentences and have to reread. Powerful.

simile. A metaphor that includes a comparative word, such as *like* or *as*.

simple sentence. A sentence that contains exactly one independent clause and no dependent clauses: *Xavier's snow blower sputtered and conked out*. Compare **complex sentence, compound-complex sentence, compound sentence**.

singular *they*. The widely condoned[186] yet still controversial use of *they* (or any of the third-person-plural pronouns—*their, them, themselves*) with

---

186. Phonetics professor Mark Liberman sums it up this way: "'Singular they' is deprecated by a few authorities, but is supported by most informed grammarians, and has often been used by great writers over the centuries" ("The SAT Fails a Grammar Test," *Language Log* blog, January 31, 2005, http://itre.cis.upenn.edu/~myl/languagelog/archives/001863.html).

a singular noun (*friend, neighbor, dentist*). In *An employee who is snowed in can't help missing their meetings*, the pronoun *their* is called a singular *they* because its antecedent, *employee*, is singular.

spliced sentence. See **comma-spliced sentence.**

squinting modifier. A word or phrase tucked confusingly between two elements, looking at both. In *Dale said tonight he'd build a fire*, the squinting modifier is *tonight*. Does *tonight* modify the verb to its left (*said*), as in "Dale made his statement tonight"? Or does it modify the verb to its right (*build*), as in "Dale will build the fire tonight"?

Possible fixes: *Tonight, Dale said he'd build a fire* or *Dale said he'd build a fire tonight*.

Compare **dangling modifier** and **misplaced modifier.**

stem sentence. A subheading traditionally used in a technical procedure to flag the break between any introductory paragraphs and the first step. Stem sentences typically start with an infinitive: *To replace the snow-blower rotator gizmo...*

structure-class words (structure words, function words). Words in any of the structure classes: prepositions (*with*), pronouns (*he*), conjunctions (*but*), determiners (*the*), auxiliaries (*might*), qualifiers (*very*), relatives (*whose*), and interrogatives (*where*).

Structure-class words have something in common that sets them apart from form-class words (parts of speech): structure-class words generally have only one form; in natural usage, they do not change form. (*The* does not appear as *thes, the's, theicity, thely, theing, unthe.*)

Unlike form-class words, structure-class words have no features of form—no defining characteristics based on form alone. In isolation (out of context), these words cannot be linguistically tested in ways that help classify them. No structure-class word—not even the words listed at the top of this definition as typical examples (*with, he, but*, etc.)—can be called a preposition in form, a pronoun in form, a conjunction in form, an *anything* in form.

Instead, structure-class words, or function words, are characterized by function: they build relationships between the form-class words around them. Words from the structure classes contain not lexical meaning (as form-class words do) but grammatical meaning; they give sentences structure and coherence. Only the most rudimentary sentences (*See Spot dig*) could exist without them.

Punctuation decisions, like sentences themselves, often hinge on structure-class words.

subordinate clause. See **dependent clause.**

subordinating conjunction (subordinator). A conjunction that subor-
dinates a clause, transforming it into a dependent clause while joining
it to an independent clause. Words that typically act as subordinating
conjunctions include *although, as if, because, unless, whenever,* and *while.*
   In *Judith sprinkles rock salt on the porch because it makes the ice melt
faster,* the word *because* is a subordinating conjunction. No punctuation
is needed between clauses when the dependent clause comes second.
When the dependent clause leads, however, you follow it with a comma:
Because rock salt makes ice melt faster, Judith sprinkles some on the porch.
   (The words identified here as subordinating conjunctions may also play
other roles, in which case they are classified differently.) Compare **conjunctive adverb** and **coordinating conjunction.**

syntax. The arrangement of words, phrases, and clauses to create well-formed
sentences—the result of the writer's answer to the question, should I put
this first or that? (or maybe, should I put this or that first?).

tense. A verb attribute that locates the action in time: past (*Bob shoveled*),
present (*Bob shovels*), future (*Bob will shovel*). Verb-tense variations, with
their sundry participles and auxiliaries, require the use of words like *plu-
perfect,* which I leave to you to sort out. Compare **aspect, mood, voice.**
See also **auxiliary.**

they, **singular.** See **singular they.**

tone. See **voice.**

topic sentence. A sentence that summarizes a paragraph's main idea. Not every
paragraph has a topic sentence. Those that do typically start with it. Sometimes, for dramatic effect, a paragraph builds to a topic sentence at the end.

*to-*verb. See **infinitive.**

transitive and intransitive verbs. These two verb types are best defined
side-by-side:
   • **Transitive verb:** A verb that has a direct object; the verb *trans*fers
      action to a noun (*trans* = "across"). For example, in *The mail carrier
      bought some fur-lined boots,* the verb *bought* is a transitive verb, and
      boots is its direct object (the noun to which it transfers action).
   • **Intransitive verb:** A verb that has no direct object. In *The mail fell
      onto the ground,* the verb *fell* is intransitive; *ground* is not a direct
      object of the verb but an object of the preposition *onto.*
   Some verbs can play either a transitive or an intransitive role. The verb
fell, for example, is intransitive in *The mail fell onto the ground* (no direct
object) and transitive in *The mail carrier is going to fell that tree* (direct
object = *tree*).

unit modifier. See **compound modifier.**

verb. To call any word a verb is ambiguous. Is it a verb in form? In function? Both?

• **Verb in form:** A form-class word (*shovel, feel, seem*) that can change form, in natural usage, in ways characteristic of verbs. In other words, a verb in form is a word that has verb features of form. In isolation, it can pass linguistic tests for verbness. *Shovel*, the standalone word, qualifies as a verb in form (example tests: *shovel+s* = third-person singular; *shovel+ing* = present participle). Of course, *shovel* also qualifies as a noun in form; like many English words, it belongs to multiple form classes.

• **Verb in function:** Any word or phrase that acts as a main verb in a phrase or clause. A verb in function typically designates actions, sensations, or states. In *Carl has been shoveling all morning*, the word *shoveling* is a verb not only in form (it ends with -*ing*) but also in function because it designates Carl's action.

Part of determining a verb's function is identifying whether it acts as a linking verb, a transitive verb, or an intransitive verb. In *Carl has been shoveling all morning*, *shoveling* plays an intransitive role because it has no direct object. Incidentally, the other two verbs—*has* and *been*—belong (by virtue of their position directly in front of the main verb) not to the form class known as verbs but to the structure class known as auxiliaries, a separate grammatical entity.

Verbs, the most complicated part of speech, have several other attributes: voice, tense, aspect, and mood.

See also **phrasal verb.**

verb particle. A word that works with a main verb, and sometimes with other words, to create a phrasal verb. For example, in the phrasal verb *chip in* ("help"), *in* is a verb particle. In the phrasal verb *drop out of* ("quit"), *out* and *of* are verb particles.

voice.

• A verb attribute that relates the subject to the action. Voice comes in two types: active and passive. Compare **aspect, mood, tense.** See also **auxiliary.**

• The writer's personality (supposedly) coming through in the words. *Voice* and *tone* are sometimes used synonymously. Those who differentiate between them say that tone is the writer's attitude toward the subject or audience in a given piece (or, more accurately, the feelings that the piece stirs in the reader, the only thing the reader can determine). A reader might perceive that a personal letter, for example, has a conciliatory tone. Tone may change from one piece of writing to another.

On the other hand, voice (as in the writer's personality) theoretically remains constant across a body of work. I don't know what to make of that claim since writers hardly remain constant across a body of work.

Ultimately, both tone and voice emerge from the writer's entire set of writing decisions—word choice, syntax, sentence length, punctuation, metaphors—everything that this book touches on and more. Base each decision on your purpose and audience. Tone and voice will follow.

• Corporate voice, the personality of an organization. An organization's writers need style guidelines to help them keep their writing consistent and on brand. If they're lucky, they also have editors—not robotic style-checking programs, which catch only the most basic of *faux pas*, but human editors, people who possess command of the language, people who know from voice.

Topics Index

If you look at only one entry in this topics index, make it "emphasis, tools for creating." And "ear, writing for the" and "eye, writing for the." Okay, that's three. Even if you never jump to the page numbers listed there, the entries alone give you information that you won't find anywhere else in the book. That's topic analysis for you. While you're at it, see "reflexivity in language about language." That's four. I draw the line at four. Four's my limit. (See "verbs," "myths," "*of.*" Don't miss "department of redundancy department.")

A note on all three indexes: Despite the pitfalls, I created this book's indexes myself.[187] Why? Indexing is the original metadata brainteaser. It gets my endorphins racing. Indexing my own work also reveals (and gives me a chance to fix) omissions, weaknesses, and inconsistencies in the text. For example, creating the entry "rhetorical devices" prompted me to identify certain rhetorical devices that I had used but hadn't thought to name or hadn't realized had a name. Take anadiplosis. Now there's a rhetorical device you can use. Look it up.

Thanks go to Jan C. Wright and Olav Martin Kvern for the Creative Commons scripts—and the inspiration and support—that enabled me to single source this book's indexes, that is, to create both e-book and print indexes from a single InDesign file.[188] Thanks also go to Scott Smiley, whose Sky Index pattern-matching tips saved me days. Anyone reading from an e-book right now owes the index links to Jan, Olav, and Scott.

May you find in these indexes, and in your life, much that you seek—and many serendipities besides.

---

187. For pros and cons of author-prepared indexes, see Mulvany, "Who Should Prepare the Index?" in *Indexing Books,* 28–34.

188. Jan C. Wright, "InDesign ePub Scripts," *Wright Information Indexing Services,* accessed November 7, 2012, http://www.wrightinformation.com/Indesign%20scripts /Indesignscripts.html.

Glossary references are in quotation marks.

: (colon). *See* colons
, (comma). *See* commas
— (dash). *See* dashes, em
! (exclamation point), 177–78, 179
- (hyphen). *See* hyphens
( ) (parentheses), 86–87
. (period). *See* periods
; (semicolon). *See* semicolons
[ ] (square brackets), 51, 86

2-by-3-inch screen template, 111 (image)

a and *an* as determiners, 198
a bit as filler phrase, 207 (under "qualifier")
about as verb particle, 13 (ftnote #26)
abstracts. *See* advance organizers
action, grammatical
   direct objects as action receivers (*See* transitive verbs)
   subjects as action performers (*See* active voice)
   subjects as action receivers (*See* passive voice)
action, narrative
   in steps, 125–26
   in storytelling, 173, 179
active voice
   "active voice," 191
   example, 173
   recommendations, 16, 17, 179
actually as filler word, 43
adaptive content, 3
adjectival compounds. *See* compound modifiers
adjectivals, 66 (image), 191
adjectives
   "adjective," 191
   definition form for, 77–78 (ftnote #96)
   as form-class words, 61–67, 200, 205 (under "modern parts of speech")
   hyphenated, 33–39

adjectives (*continued*)
   recommendations, 18, 67
   sentence diagram, 66 (image)
   verbs in role of, 44
advance organizers, 115 (image), 124–25, 192
"adverbial," 192
adverbs
   "adverb," 192
   as form-class words, 61–67, 200, 205 (under "modern parts of speech")
   prepositional phrases in role of, 206 (under "preposition")
   recommendations
     placing accurately, 23
     using creatively, 44, 67, 175
     using sparingly, 18, 43, 175
     using without hyphens, 35
   vs. prepositions and verb particles, 49–52, 55
ad writing, 110
"Alcohol! I only drink..." sign, 21 (image)
all-encompassing vs. *all encompassing*, 37
alliteration, 134, 172, 192
along as adverb or verb particle, 55
although as subordinating conjunction, 211
"Always ___" (fill in the blank). *See* myths
am. See *be*-verbs
Amazon.com
   nontester of user-guide text, 119–22
   tester of website text, 89–90
"amplification," 193
anadiplosis, 169–70, 193
analogy as flawed method of reasoning, 7, 193. *See also* metaphors
"anaphora," 193
and
   complexity of classifying, 64
   as coordinating conjunction, 91, 197
   as sentence starter, 197 (under "coordinating conjunction")

Names Index

This index (like the other two, per common practice) ignores most front and back matter and all epigraphs, those quotations at the beginning of each chapter. So if you're looking for Aristotle, Doonesbury, or my mom, you have to riffle. Otherwise, if the book mentions or cites a person, persona, character, pseudonym, or British comedy group, that name should show up here.

Titles Index

This index includes titles of books, articles, and blog posts mentioned or cited. It also includes the odd podcast series, presentation, radio show, speech, or sketch by a British comedy group. If you're looking for a title and come up dry, you're my kind of reader. You'll have scribbled a note on that page and won't have a lick of trouble finding it again.

CPSIA information can be obtained at www.ICGtesting.com
Printed in the USA
BVOW03s1056021213

337903BV00008B/160/P